Advanced Structural Materials—2011

MATERIALS RESEARCH SOCIETY
SYMPOSIUM PROCEEDINGS VOLUME 1373

Advanced Structural Materials—2011

EDITORS

Hector A. Calderon
Departamento de Física, ESFM-IPN
México D.F., México

Armando Salinas Rodriguez
CINVESTAV-Unidad Saltillo
Coahuila, México

Heberto Balmori Ramirez
ESIQIE-IPN
México D.F., México

MRS

Materials Research Society
Warrendale, Pennsylvania

CAMBRIDGE
UNIVERSITY PRESS

CAMBRIDGE UNIVERSITY PRESS
Cambridge, New York, Melbourne, Madrid, Cape Town,
Singapore, São Paulo, Delhi, Mexico City

Cambridge University Press
32 Avenue of the Americas, New York, NY 10013-2473, USA

www.cambridge.org
Information on this title: www.cambridge.org/9781605113500

Materials Research Society
506 Keystone Drive, Warrendale, PA 15086
http://www.mrs.org

First published 2012

CODEN: MRSPDH

ISBN: 978-1-60511-350-0 Hardback

CONTENTS

PREFACE

This volume is a compilation of papers that were presented at Symposium 4 of the XX International Conference of the Mexican Academy of Materials Science – MRS Mexico, organized in collaboration with MRS, "Advanced Structural Materials" which was held from August 14th to 19th, 2011 in Cancun, Mexico. The symposium was devoted to fundamental and technological applications of structural materials, and continued the tradition of providing a forum for scientists from various backgrounds with a common interest in the development and use of structural materials to come together and share their findings and expertise.

The papers contained in this volume are a collection of invited and contributed papers. This year, the symposium was attended by participants from Argentina, Belgium, Brazil, Czech Republic, France, Japan, Mexico, Poland, Turkey, Spain, Switzerland and the United States. All papers have been thoroughly reviewed by at least two referees and edited to the standards of the Materials Research Society. We are grateful to all those referees who, by their comments and constructive criticism, helped to improve the final printed papers, and to all the authors who made additional efforts to prepare their manuscripts

The Advanced Structural Materials symposium has been held for the last 14 years with the objective of presenting overviews and recent investigations related to advanced structural metallic, ceramic and composite materials. The topics include innovative processing, phase transformations, mechanical properties, oxidation resistance, modeling and the relationship between processing, microstructure and mechanical behavior. Additionally, papers on industrial application and metrology are included.

The organizing committee gratefully acknowledges the enthusiastic cooperation of all symposium participants, as well as the kind acceptance of the editorial committee of Materials Research Society to publish these proceedings. The financial support of the Instituto Politécnico Nacional (IPN, Mexico) and the Centro de Investigación y de Estudios Avanzados del Instituto Politécnico Nacional (CINVESTAV-IPN) is also acknowledged. We hope that all readers will come to consider the Advanced Structural Materials Symposium at Cancun as a suitable forum to present results of their recent research and experience.

Greetings from the Organizing Committee

The Organizing Committee:

Heberto Balmori-Ramírez (Instituto Politécnico Nacional).
Hector A. Calderon-Benavides (Instituto Politécnico Nacional).
Armando Salinas-Rodríguez (Centro de Investigación y de Estudios Avanzados del IPN).

Materials Research Society Symposium Proceedings

Volume 1371 — Nanostructured Materials and Nanotechnology, C. Gutiérrez-Wing, J.L. Rodríguez-López, O.A. Graeve, J.J. Boeckl, P. Soukiassian, 2012, ISBN 978-1-60511-348-7
Volume 1372 — Structural and Chemical Characterization of Metals, Alloys, and Compounds – 2011, R. Pérez Campos, A. Contreras Cuevas, R.A. Esparza Munoz, 2012, ISBN 978-1-60511-349-4
Volume 1373 — Advanced Structural Materials – 2011, H.A. Calderon, A. Salinas Rodriguez, H. Balmori Ramirez, 2012, ISBN 978-1-60511-350-0
Volume 1374 — Cultural Heritage and Archaeological Issues in Materials Science, J.L. Ruvalcaba Sil, J. Reyes Trujeque, A.Velazquez Castro, M. Espinosa Pesqueira. 2012, ISBN 978-1-60511-351-7
Volume 1376E — Biomaterials for Medical Applications, S. Rodil, A. Almaguer, K. Anselme, 2012, ISBN 978-1-60511-353-1
Volume 1380E — Materials Research for Mining and Mineral Processing, F.R.C. Pedroza, 2012, ISBN 978-1-60511-357-9
Volume 1381E — Materials Welding and Joining Technologies, F.A.R. Valdes, 2012, ISBN 978-1-60511-358-6
Volume 1383 — Material Challenges in Current and Future Nuclear Technologies, K.R. Whittle, M. Bertolus, B. Uberuaga, R.W. Grimes, 2011, ISBN 978-1-60511-360-9
Volume 1384E — Advanced Materials for Fuel Cells, J. Hertz, M.L. DiVona, P. Knauth, H.L. Tuller, 2011, ISBN 978-1-60511-361-6
Volume 1385E — *In-Situ* Studies of Solid-Oxide Fuel-Cell Materials, R. Maher, 2011, ISBN 978-1-60511-362-3
Volume 1386E — Sustainable Synthesis of Nanomaterials, H. Fan, M. Knez, S.S. Wong, W. Lee, 2011, ISBN 978-1-60511-363-0
Volume 1387E — Advanced Materials for Solar-Fuel Generation, C. Hill, 2011, ISBN 978-1-60511-364-7
Volume 1388 — Mobile Energy, S. Mhaisalkar, K. Shenai, G. Amaratunga, A. Nathan, 2011, ISBN 978-1-60511-365-4
Volume 1389E — Applications of Hierarchical 3D Structures, J.H. Moon, S. Jeon, S. Yang, R.A. Vaia, 2011, ISBN 978-1-60511-366-1
Volume 1390 — Organic Photovoltaics–Materials to Devices, V. Bommisetty, G. Li, C. Deibel, T-Q. Nguyen, D.C. Olson, M. Riede, M. Leclerc, V. Dyakonov, G. Rumbles, N.S. Sariciftci, 2011, ISBN 978-1-60511-367-8
Volume 1391E — Photonic and Plasmonic Materials for Enhanced Photovoltaic Performance, R. Biswas, 2011, ISBN 978-1-60511-368-5
Volume 1392E — Materials for High-Performance Photonics, T.M. Cooper, S.R. Flom, M. Bockstaller, C. Lopes, 2011, ISBN 978-1-60511-369-2
Volume 1393E — Topological Insulator Materials, C. Felser, Y. Cui, H. Peng, S. Murakami, 2011, ISBN 978-1-60511-370-8
Volume 1394E — Oxide Semiconductors–Defects, Growth and Device Fabrication, T. Veal, S. Durbin, J. Phillips, M. Grundmann, 2011, ISBN 978-1-60511-371-5
Volume 1395 — Diamond Electronics and Biotechnology–Fundamentals to Applications V, O.A. Williams, R.B. Jackman, P. Bergonzo, G.M. Swain, K.P. Loh, 2011, ISBN 978-1-60511-372-2
Volume 1396 — Compound Semiconductors for Generating, Emitting and Manipulating Energy, T. Li, M. Mastro, A. Dadgar, H. Jiang, J. Kim, 2011, ISBN 978-1-60511-373-9
Volume 1397E — Ferroelectric and Multiferroic Materials, M. Bibes, C.J. Fennie, L.W. Martin, B. Noheda, T. Kimura, 2011, ISBN 978-1-60511-374-6
Volume 1398E — Magnetoelectric Composites, P. Finkel, 2011, ISBN 978-1-60511-375-3
Volume 1399E — Compliant Electronics and Photonics, D. Tyler, 2011, ISBN 978-1-60511-376-0
Volume 1400E — Solution Processing of Inorganic and Hybrid Materials for Electronics and Photonics, P.J. Smith, M.F.A.M. van Hest, D.B. Mitzi, A. Morrin, 2011, ISBN 978-1-60511-377-7
Volume 1401E — Large-Area Processing and Patterning for Active Optical and Electronic Devices III, I. Kymissis, T. Anthopoulos, C. Madigan, M. Shtein, 2011, ISBN 978-1-60511-378-4
Volume 1402E — Charge Generation/Transport in Organic Semiconductor Materials, J. Anthony, 2011, ISBN 978-1-60511-379-1
Volume 1403 — Multifunctional Polymer-Based Materials, A. Lendlein, Y. Feng, T. Xie, Z. Guan, 2011, ISBN 978-1-60511-380-7
Volume 1404E — Phonons in Nanomaterials–Theory, Experiments and Applications, S.L. Shinde, D.H. Hurley, G.P. Srivastava, M. Yamaguchi, 2011, ISBN 978-1-60511-381-4

MATERIALS RESEARCH SOCIETY SYMPOSIUM PROCEEDINGS

Volume 1405E — Advances in Energetic Materials Research, M.R. Manaa, C-S. Yoo, E.J. Reed, M.S. Strano, 2011, ISBN 978-1-60511-382-1
Volume 1406 — Functional Metal-Oxide Nanostructures, A. Vomiero, S. Mathur, Z.L. Wang, E. W-G. Diau, 2011, ISBN 978-1-60511-383-8
Volume 1407 — Carbon Nanotubes, Graphene and Related Nanostructures, Y.K. Yap, D. Futaba, A. Loiseau, M. Zheng, 2011, ISBN 978-1-60511-384-5
Volume 1408 — Functional Nanowires and Nanotubes, K. Nielsch, A.F. i Morral, H. Linke, H. Shin, L. Shi, 2011, ISBN 978-1-60511-385-2
Volume 1409E — Functional Semiconductor Nanocrystals and Metal-Hybrid Structures, K.S Leschkies, P. Nagpal, M.A. Pelton, H. Mattoussi, P. Kambhampati, 2011, ISBN 978-1-60511-386-9
Volume 1410E — Transport Properties in Polymer Nanocomposites II, S. Nazarenko, J. Grunlan, J. Bahr, E. Espuche, 2011, ISBN 978-1-60511-387-6
Volume 1411E — Self Organization and Nanoscale Pattern Formation, S. Persheyev, 2011, ISBN 978-1-60511-388-3
Volume 1412E — Mechanical Nanofabrication, Nanopatterning and Nanoassembly, G. Cross, A. Schirmeisen, A. Knoll, M. Rolandi, 2011, ISBN 978-1-60511-389-0
Volume 1413E — Safety and Toxicity Control of Nanomaterials, W.W. Yu, V.L. Colvin, Q. Dai, P.C. Howard, 2011, ISBN 978-1-60511-390-6
Volume 1415 — MEMS, BioMEMS and Bioelectronics–Materials and Devices, T. Albrecht, M.P. de Boer, F.W. DelRio, M.R. Dokmeci, C. Eberl, J. Fukuda, H. Kaji, C. Keimel, A. Khademhosseini, 2011, ISBN 978-1-60511-392-0
Volume 1416E — Nanofunctional Materials, Nanostructures and Nanodevices for Cancer Applications, S. Svenson, P. Grodzinski, S. Manalis, X J. Liang, W. Lin, 2011, ISBN 978-1-60511-393-7
Volume 1417E — Biomaterials for Tissue Regeneration, C.C. Sorrell, 2011, ISBN 978-1-60511-394-4
Volume 1418 — Gels and Biomedical Materials, F. Horkay, R. Narayan, V. Dave, S. Jin, N. Langrana, J.D. Londono, W. Oppermann, S. Ramakrisha, D. Shi, R.G. Weiss, 2011, ISBN 978-1-60511-395-1
Volume 1419E — Nucleation and Growth of Biological and Biomimetic Materials, P.M. Rodger, J. Harding, L.B. Gower, P. Vekilov, 2011, ISBN 978-1-60511-396-8
Volume 1420E — Multiscale Mechanics of Hierarchical Materials, F. Barthelat, 2011, ISBN 978-1-60511-397-5
Volume 1421E — Three-Dimensional Tomography of Materials, S. Pennycook, 2011, ISBN 978-1-60511-398-2
Volume 1422E — Functional Imaging of Materials–Advances in Multifrequency and Multispectral Scanning Probe Microscopy and Analysis, A. Baddorf, 2011, ISBN 978-1-60511-399-9
Volume 1423E — Dynamics in Confined Systems and Functional Interfaces, M.H. Müser, D.L. Irving, S.B. Sinnott, I. Szlufarska, 2011, ISBN 978-1-60511-400-2
Volume 1424 — Properties and Processes at the Nanoscale–Nanomechanics of Material Behavior, D. Bahr, P. Anderson, N. Moody, R. Spolenak, 2011, ISBN 978-1-60511-401-9
Volume 1425E — Combinatorial and High-Throughput Methods in Materials Science, J.B. Miller, J. Genzer, Y. Matsumoto, R.A. Potyrailo, 2011, ISBN 978-1-60511-402-6

Prior Materials Research Society Symposium Proceedings available by contacting Materials Research Society

Mater. Res. Soc. Symp. Proc. Vol. 1373 © 2012 Materials Research Society
DOI: 10.1557/opl.2012.288

Structure of Dislocations and Mechanical Properties of B2 Alloys

V. Paidar[1] and V. Vitek[2]
[1]Institute of Physics AS CR, Na Slovance 2, 182 21 Praha 8, Czech Republic
[2]Department of Materials Science and Engineering, University of Pennsylvania,
3231 Walnut Street, Philadelphia, PA 19104, U.S.A

ABSTRACT

Adequate mechanical properties are important for both structural and functional applications of materials. There are significant differences in mechanical behaviour of different B2 ordered alloys and these are related to the properties of superlattice dislocations. Several types of dislocations can be activated, in particular <111> and <001> dislocations gliding on {110} planes. Their mobility can vary markedly from material to material and this has a strong impact on the mechanical properties. With the aim to elucidate qualitatively the differences between different alloys crystallising in the same B2 structure we analyse possible dislocation dissociations. The model employed is based on the isotropic elasticity but includes an important characteristic of stacking-fault-like defects involved in the splittings, the deviation of their displacements away from the usually assumed ½<111> APB.

INTRODUCTION

Mechanical properties of B2 intermetallic compounds and alloys are controlled by the type of active dislocations. The shortest lattice vector in the B2 structure is <001> and therefore these dislocations are likely to possess the lowest energy. Hence, the <001> slip direction is expected to be dominant. There are, indeed, B2 alloys in which the principal slip direction is <001> (e. g. CoTi [1-3]) but the <111> slip dominates in others (e.g CuZn [4] or FeAl [5, 6]) in spite of the fact that the square of the <111> Burgers vector is three times larger than that of the <001> vector and, therefore, the energy of the <111> dislocations is expected to be much higher than that of <001> dislocations. However, the dislocation energy is not the only criterion determining which dislocations mediate the slip. Another factor, which may be even more important than the energy, is the dislocation mobility that depends crucially on the dislocation core structure [7], in particular on the ability of dislocations to split into partial dislocations separated by metastable stacking-fault-like defects. For example, if dislocations possess narrow non-planar cores their motion is restricted since these cores must be first transformed from their sessile form to a glissile form. The well-known example is BCC metals. On the other hand, dislocations with planar cores are usually quite mobile since their cores are already in the glissile form. The latter is usually the case if dislocations dissociate into a crystal plane like, for example, in FCC metals.

It is commonly accepted that in materials with metastable stacking faults that possess sufficiently low energies the dislocations dissociate into well-separated partials which renders dislocation cores planar. It has been generally assumed that in B2 ordered alloys the APB of the type ½<111> is a metastable fault on {101} planes and that its energy is related to the order-disorder transition temperature between the ordered B2 and disordered BCC phase. If this were the case then when the order-disorder transition temperature is low the APB energy is low and thus <111> superdislocations may dissociate into two well separated ½<111> superpartials and in this case the <111> slip direction will be favoured [8-10]. This appears to be the case in CuZn but for example CoTi and FeAl are both ordered up to the melting temperature and as a result have very high transition temperatures but the <001> slip is dominant in the former while the <111> slip operates in the latter.

However, as found in a number of atomistic studies [11-16], the ½<111> APB may not be a metastable fault on {110} planes. Instead, displacements leading to metastable faults are not determined by crystallography but by details of atomic interactions and vary from material to material. Specifically, in the (110) plane the fault vector may be shifted away from $\frac{1}{2}[1\bar{1}1]$ along the $[1\bar{1}0]$ direction. The dislocations with the Burgers vector $\mathbf{b} = [1\bar{1}1]$ may then dissociate into three partials, as shown in Fig. 1,

$$\mathbf{b} = \mathbf{b}_1 + \mathbf{b}_2 + \mathbf{b}_1. \tag{1}$$

Similarly, the dislocation with the Burgers vector $\mathbf{a} = [001]$ may also dissociate following the reaction

$$\mathbf{a} = \mathbf{b}_1 + \mathbf{b}_3 \tag{2}$$

where $\mathbf{b}_3$ has the same magnitude as $\mathbf{b}_1$ but different direction (see Fig. 1).

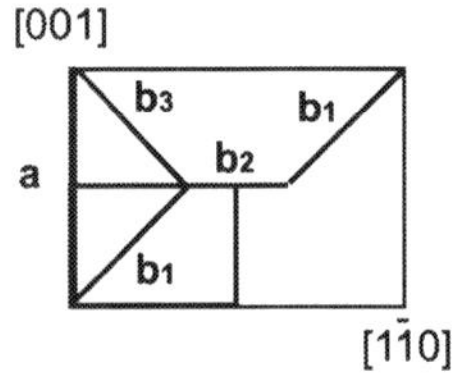

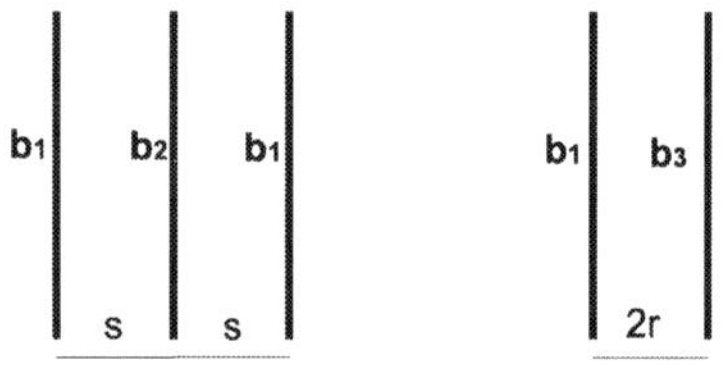

Figure 1. Dissociations of $[1\bar{1}1]$ and [001] dislocations on the (110) plane in a B2 lattice according to equations (1) and (2), respectively.

Figure 2. Definition of the widths of splittings for the $[1\bar{1}1]$ dislocation dissociated according to equation (1) (left-hand side) and the [001] dislocation dissociated according to equation (2) (right-hand), respectively.

Owing to the symmetry of the (110) plane the direction of $\mathbf{b}_2$ has to be $[1\bar{1}0]$ [17] but the direction of $\mathbf{b}_1$ varies from material to material. The vectors $\mathbf{b}_1$ (and thus also $\mathbf{b}_2$ and $\mathbf{b}_3$) are determined by minima of the γ-surface (see e. g. [7]) that has to be calculated for a given alloy using a reliable description of interatomic forces for the specific material; this is at present provided within the density functional theory (DFT) based methods while results of potentials with an empirical input are merely qualitative. In this paper we examine only the potential consequences of such non-crystallographic metastable planar faults and the corresponding character of dislocation splitting while the results for specific alloys will be published elsewhere. In this context we discuss the basic question why dislocations with much longer Burgers vector <111> rather than <001> may mediate the plastic deformation in some B2 systems. For simplicity the isotropic elasticity is employed in the following study of dislocation dissociations. While this analysis is only qualitative, it demonstrates how the type of dislocations mediating plastic flow in B2 alloys may vary with the form of metastable stacking-fault-like defects and elastic properties of the material.

ANALYSIS OF DISLOCATION DISSOCIATIONS

The most important characteristics of a stacking-fault-like planar defect on the (110) plane are the magnitude of the vector $\mathbf{b_2}$ that measures the deviation of the displacement vector from the crystallographic vector $\frac{1}{2}[1\bar{1}1]$ and its energy γ. Let us consider the whole range of such displacements, $\mathbf{b}_2 = x/2[1\bar{1}0]$, where the parameter x varies from 0, for the APB displacement, up to 1, for the displacement ½[001]. We will now analyse the dislocation splitting as a function of two parameters, x, characterising the geometry of the Burgers vectors of partial dislocations, and the quantity $M = \mu a / 4\pi\gamma$ that characterises the material via the shear modulus μ, the lattice constant a and the energy of the metastable stacking-fault-like defect on the (110) plane γ.

The widths of dislocations dissociated according to (1) and (2) are defined by separations, $2s$ and $2r$, respectively, as shown in Fig. 2. In the following we consider only screw orientations of the dislocations **b** and **a** that have lower elastic energy than corresponding edge dislocations with the same Burgers vectors and may have non-planar cores if not dissociated in the (110) plane. In the framework of linear isotropic elasticity [18] the half widths of the dissociations described by equations (1) and (2) are in the units of the lattice constant a

$$\frac{r}{a} = M\frac{1}{2}\left(\frac{1}{2} - \frac{(1-x)^2}{1-\nu}\right) \tag{3}$$

and

$$\frac{s}{a} = M\left(\frac{b_1}{a}2\sqrt{3}\cos\varphi - 3\left(\frac{b_1}{a}\right)^2\left(\cos^2\varphi + \frac{\sin^2\varphi}{1-\nu}\right)\right) \tag{4}$$

where ν is the Poisson ratio, φ the angle between the vector $\mathbf{b_1}$ and the direction $[1\bar{1}1]$ and b_1 the magnitude of the vector $\mathbf{b_1}$.

The $[1\bar{1}1]$ and [001] dislocations can be considered as dissociated if the values of r or s are at least positive. Otherwise the dislocation core is compact and possibly spread spatially. Following eqs. (3) and (4) signs of s and r are controlled by the parameter x and the Poisson ratio ν. Let us consider the Poisson ratio to be in the interval 0.2-0.5 that covers a large spectrum of B2 systems. Fig. 3 then shows the range of x and ν for which the splitting of [001] and $[1\bar{1}1]$ screw dislocations may occur. In this figure ν is depicted on the abscissa and x on the ordinate. The dissociation of the [001] dislocation can take place in the region above the lower line and the dissociation of the $[1\bar{1}1]$ dislocation below the upper line. It is seen that the [001] screw dislocation can dissociate according to equation (2) only for x sufficiently large, above about 0.4, i.e. in the upper part of the x-ν domain in Fig. 3. On the other hand, $[1\bar{1}1]$ screw dislocations may dissociate practically in the whole x-ν domain. In both cases the width of the dissociation depends, of course, on the value of M.

The difference between s and r, i.e. the difference of the half widths of the splittings of $[1\bar{1}1]$ and [001] dislocations, is plotted in Fig. 4 as a contour map showing the lines of constant differences of separations of partial dislocations. It becomes negative in the upper-right corner of this diagram only when the width of splitting of the [001] dislocation is larger than that of the $[1\bar{1}1]$ dislocation. The difference is positive and large for low values of the deviation x when the dissociation of the $[1\bar{1}1]$ dislocation will be much wider.

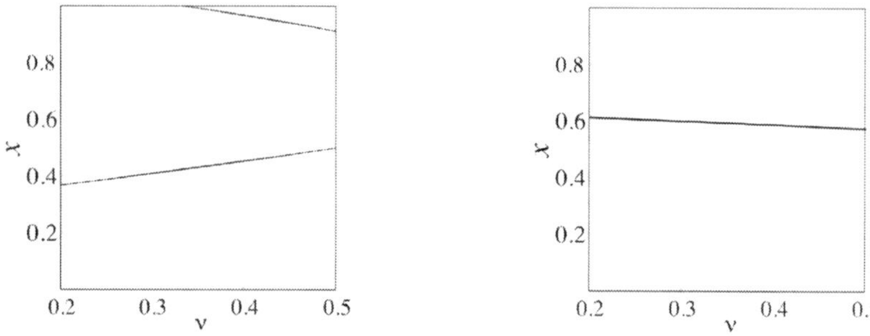

Figure 3. Domains of x and ν in which $[1\bar{1}1]$ and [001] screw dislocations may dissociate according to equations (1) and (2), respectively. The [001] dislocation may split in the region above the lower line and the $[1\bar{1}1]$ dislocation below the upper line.

Figure 4. The differences in the width of splittings between the $[1\bar{1}1]$ and [001] screw dislocations depicted in x - ν plot. Below the line the width of splitting of the $[1\bar{1}1]$ screw dislocation is by more than $2a$ larger than that of the [001] screw dislocation.

As mentioned in the introduction, the widely dissociated dislocations are anticipated to be more mobile than the compact ones that may have non-planar cores when in screw orientation. Let us assume that if the width of the splitting of the $[1\bar{1}1]$ screw dislocation is wider than that of the [001] dislocation by about $2a$ it is more mobile. In Fig. 4 ν is again depicted on the abscissa and x on the ordinate. In the region below the line the width of splitting of the $[1\bar{1}1]$ screw dislocation, s, is more than $2a$ larger than that of the [001] screw dislocation, r, and thus in this region of x and ν the $[1\bar{1}1]$ screw dislocation is very likely to be more mobile. It is seen from Fig, 5 that this is possible only if $x \leq 0.6$. This is, of course, only a qualitative consideration of the dependence of the width of splitting and thus possibly dislocation mobility on the deviation of the fault vector away from the usually assumed $\frac{1}{2}[1\bar{1}1]$ APB that will be different for different materials. Moreover, the width of splitting will also depend on the amount of elastic anisotropy. Nevertheless, the present example demonstrates how the differences in mobility of <111> and <001> dislocations may depend on the fact that the metastable planar faults on {110} planes are not necessarily the usually assumed ½<111> APBs.

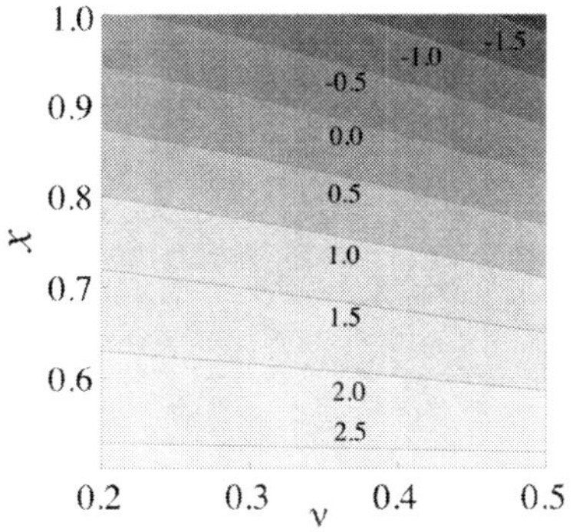

Figure 5. Contour map of constant differences between the half widths *s* and *r* of the splittings of $[1\bar{1}1]$ and [001] screw dislocations. The notation of the contours of the constant difference between *r* and *s* is in units of the lattice parameter *a*.

DISCUSSION

Assuming that the mobility of screw dislocations in B2 alloys depends on their ability to dissociate on {110} planes, we have investigated how this depends on the deviation of the displacements of metastable faults on these planes from the usually assumed vector ½<111>, associated with the {110} APB. This deviation is characterised by the vector $x/2<110>$, where $0 \le x \le 1$. Notice that for $x = 1$ the position of the local minimum of the stacking-fault-like defect would be at the point ½<001> but here any γ-surface exhibits a local maximum [19]. Therefore, very large deviations of the displacement vector away from ½<111> are not likely and thus the region in which the width of splitting of <001> dislocations would be larger that that of <111> dislocations is rather limited (see Fig. 4).

Notwithstanding, in specific materials the widths of dislocation splitting (cores) depend also on material properties enclosed in the present study in the parameter *M*. When it is large, for example because of the low energy of the stacking-fault-like defect, as in CuZn, <111> dislocations will be significantly more split than <001> dislocations, which may not be dissociated at all. On the other hand, when *M* is small, as for example in FeTi, <111> dislocations will be virtually undissociated for any value of *x* and <001> dislocations, albeit also undissociated, will dominate energetically.

In spite of the fact that our analysis provides only a qualitative information on the dissociation of dislocations in B2 alloys, it clearly demonstrates that it is essential to take into account the fact that partial dislocations involved in the splitting of dislocations in these alloys may have Burgers vectors significantly different from the usually assumed ½<111>. A more precise description of dislocation dissociations in specific B2 alloys has to incorporate over and above the considerations of the displacement vectors of stacking-fault-like defects and their energies, the influence of the elastic anisotropy and this will be presented in a forthcoming paper.

CONCLUSIONS

A generic model has been presented that illustrates the effect of the deviation of displacements characterising the metastable planar faults on {101} planes, away from the commonly assumed ½<111> APB, on dislocation behaviour and thus on mechanical properties of B2 intermetallics. The primary consequence is that instead of the two-fold splitting of <111> dislocations, these dissociation may be three-fold, i. e. into three partial dislocations. It has been shown that <001> dislocations are likely to be more widely split than <111> only for large deviations of the fault vectors from ½<111>. However, elastic moduli and energy of the planar faults determine how wide the splitting of <111> dislocations is. In the framework of isotropic elasticity this is controlled by the parameter $M = \mu a / 4\pi\gamma$ that is small for large energies γ. Hence, if the deviation of the fault vector away from ½<111> is not large and the fault energy γ is small enough the <111> dislocations are widely split and dominate the plastic deformation. The <001> dislocations may dominate either if the deviation of the fault vector from ½<111> is large or when this deviation is not too large but the fault energy γ is so large that the <111> dislocation is undissociated and thus very sessile in the screw orientation (see e. g. [14]). It is, of course, possible that in this case the <001> dislocations are also highly sessile and then the material is brittle since all the screw dislocations are immobile.

ACKNOWLEDGMENTS

This research was supported in part (VP) by the Grant Agency of the Academy of Sciences of the Czech Republic, contract No. IAA100100920, and in part (VV) by the US Department of Energy, BES, Grant no. DEPG02-98ER45702.

REFERENCES

1. T. Takasugi, K. Tsurisaki, O. Izumi and S. Ono, Phil. Mag. A **61** (1990) 785.
2. D. Shindo, M. Yoshida, B. T. Lee, T. Takasugi and K. Hiraga, Intermetallics **3** (1995) 167.
3. J. A. Wollmershauser, C. J. Neil and S. R. Agnew, Met. Mater. Trans. A **41** (2010) 1217.
4. Y. Umakoshi, M. Yamaguchi, Y. Namba and M. Murakami, Acta Metall. **24** (1976) 89.
5. T. Yamagata and H. Yoshida, Mater. Sci. Eng. **12** (1973) 95.
6. D. Wu, I. Baker, P. R. Munroe and E. P. George, Intermetallics **15** (2007) 103.
7. V. Vitek and V. Paidar, in "Dislocations in Solids", edited by J. P. Hirth (North-Holland, Amsterdam, 2008) p. 439.
8. W. A. Rachinger and A. H. Cottrell, Acta Metall. **4** (1956) 109.
9. M. H. Yoo, T. Takasugi, S. Hanada and O. Izumi, Mater. Trans. JIM **31** (1990) 435.
10. T. Takasugi, S. Hanada, M. Yoshida and D. Shindo, Phil. Mag. A **71** (1995) 347.
11. T. A. Parthasarathy, S. I. Rao and D. M. Dimiduk, Phil. Mag. A **67** (1993) 643.
12. D. Farkas, S. J. Zhou, C. Vaihle, B. Mautasa and J. Panova, J. Mater. Res. **12** (1997) 93.
13. C. Vailhe and D. Farkas, Acta Mater. **45** (1997) 4463.
14. R. Schroll, V. Vitek and P. Gumbsch, Acta Mater. **46** (1998) 903.
15. C. Vailhe and D. Farkas, Phil. Mag. A **79** (1999) 921.
16. V. Paidar, Y.-S. Lin, M. Cak and V. Vitek, Intermetallics **18** (2010) 1285.
17. M. Yamaguchi, D. P. Pope, V. Vitek and Y. Umakoshi, Phil. Mag. A **43** (1981) 1265.
18. J. P. Hirth and J. Lothe, Theory of Dislocations (Pergamon, Oxford, 1982).
19. V. Paidar and V. Vitek, in "Intermetalic Compounds, Progress", edited by J. H. Westbrook and R. L. Fleischer (John Wiley, Chichester, 2002) p. 437.

Mater. Res. Soc. Symp. Proc. Vol. 1373 © 2012 Materials Research Society
DOI: 10.1557/opl.2012.289

Exploring the Synthesis Parameters and Spark Plasma Sintering of Tantalum Carbide Powders Prepared by Solvothermal Synthesis

Braeden M. Clark, James P. Kelly, and Olivia A. Graeve*
Kazuo Inamori School of Engineering, Alfred University, Alfred, NY 14802

*Author to whom correspondence should be addressed: Email: graeve@alfred.edu, Tel: (607) 871-2749, Fax: (607) 871-2354, URL: http://people.alfred.edu/~graeve/

ABSTRACT

Tantalum carbide is a technologically important material for use in ultra-high temperatures and corrosive environments. In this report, we describe the scalability of a low temperature solvothermal method for the preparation of this useful material. X-ray diffraction shows phase-pure powders with no change in average crystallite size or compound stoichiometry compared to synthesis in smaller batches, remaining at 25 nm and 0.94 respectively. Dynamic light scattering shows a slight decrease in particle size distribution with scale-up. Thermogravimetric analysis (TGA) in air shows a decrease in surface species on the powders, but the powders oxidize at a lower temperature when scaling the synthesis. Mass spectrometry performed alongside TGA in a helium atmosphere reveals that water is the most abundant species on the surface of the powders, but oxygen, carbon monoxide, carbon dioxide, and nitrogen are also detected. Oxygen analysis reveals that the oxygen content of the powders is high (>6%). The oxygen source and methods of decreasing oxygen content are discussed. Initial sintering trials were performed and demonstrate the need for further powder processing.

INTRODUCTION

Tantalum carbide is of special interest because of its high melting temperature (~4000 K), making it one of the most refractory ceramics known. In addition to being refractory, other properties have been investigated including mechanical strength, oxidation behavior, elasticity, superconductivity, electronic structure, thermodynamic properties and hardness [1-8]. Due to these properties, tantalum carbide is being investigated as a candidate material for use in space reentry vehicles, high Mach aircraft, and propulsion systems, where high temperatures and pressures are sustained.

A novel solvothermal synthesis method has been used to successfully synthesize tantalum carbide and other non-oxide ceramic powders [9-10]. This technique makes use of fused-quartz tubes to allow observation of the reaction; however, this limits the amount of powder that can be synthesized per reaction (~3 g) and risks contamination from the reaction vessel. A scale-up synthesis using the same principles is done in a stainless-steel beaker to reduce contamination and demonstrate scalability of this process.

EXPERIMENTAL PROCEDURE

Reactants for the synthesis were prepared in an argon-filled glove box. The precursors consisted of 92.8 g tantalum (V) chloride (99.8%, Alfa Aesar, Ward Hill, MA) and 9.3 g of carbon (Lampblack 101, Degussa, Parsippany, NJ) and were mixed with a mortar and pestle.

The powders were added to a 600 mL stainless-steel beaker along with 27 g of lithium (granular, 99%, Sigma-Aldrich, St. Louis, MO) and covered with a stainless-steel pan with a weight on top. The reactants were removed from the glove box and placed on a hot-plate inside a chemical fume hood. The hot plate was set at a temperature of 548 K to initiate the self-propagating reaction.

After ignition, the powders were allowed to air cool. The powders were then removed from the beaker by repeatedly adding enough deionized water to submerge the powders and quench the reaction with the excess lithium. The powders and water were added to a beaker and water was added to form a suspension of 1000 mL. The suspension was magnetically stirred for 30 minutes, ultrasonicated for 30 minutes, magnetically stirred for another 15 minutes, centrifuged at 11,000 rpm for 5 minutes, and decanted before washing of the powders.

The washing procedure for the TaC powders consisted of magnetically stirring for 30 minutes, ultrasonicating for 30 minutes, magnetically stirring for another 15 minutes, and centrifuging at 11,000 rpm for enough time to separate the supernatant and powders. The powders were subjected to six of these washes to eliminate excess lithium and then allowed to dry.

The cleaned powders were characterized using X-ray diffraction on a Siemens D5000 instrument scanning from 15° to 85° 2θ with a step size of 0.04° 2θ and a hold time of 1 s. Average crystallite size was determined using the Williamson-Hall technique with Jade 8 software. The lattice parameter was also found, using this software, through unit-cell refinement. The lattice parameters were used to find the compound stoichiometry using the lattice parameter and the relationship C/Ta = 6.398a – 17.516, where a is the lattice parameter [11]. Dynamic light scattering (DLS) via a Nanotrac ULTRA instrument was used to determine the particle size distribution of the powders. The samples were prepared by magnetically stirring 0.01 g of powder in 25 mL of deionized water for 24 hours and then ultrasonicating for 5 minutes before the measurements were taken. Thermogravemetric analysis (TGA) was performed using TA Instruments Q-50 TGA instrument. Weight loss as a function of temperature was measured using a heating rate of 10 K per minute to 1273 K. X-ray photoelectron microscopy (XPS) was performed using a PHI Quantera SXM system from Physical Electronics Inc. Oxygen analysis was performed using a TC600 instrument. Results are compared to a 3 g batch made according to previous work [9-10].

Spark plasma sintering (SPS) was used to sinter TaC specimens on a FCT Systeme D-25 system. Sintering was performed on 6 g samples using a 50 MPa uniaxial pressure and a 100 K/min heating rate to either 1525°C or 1725°C, after which the pressure and current were dismissed and the samples were allowed to cool as quickly as possible. The diameter of the sintered specimens was 1 inch and the thickness was 2-3 mm.

RESULTS AND DISCUSSION

Figure 1 compares the X-ray diffraction patterns of a 3 g batch and a 50 g batch of TaC before and after heat treatment. The results show phase-pure TaC for the 3 g batch whereas a small amount of tantalum metal is observed in the as-synthesized 50 g batch. The tantalum metal reacted after subjecting the powders to heat treatment and was not observed after the heat treatment. The average crystallite size and compound stoichiometry do not show significant changes in the as-synthesized state and are not affected by heat treatments up to 773 K. The average crystallite size was 25 nm for the as-synthesized powders and 26 nm after heat treatment. The compound stoichiometry was 0.94 C/Ta for powders in the as-synthesized state and 0.94 and 0.93 for the 50 g and 3 g batches, respectively, after heat treatment at 773 K.

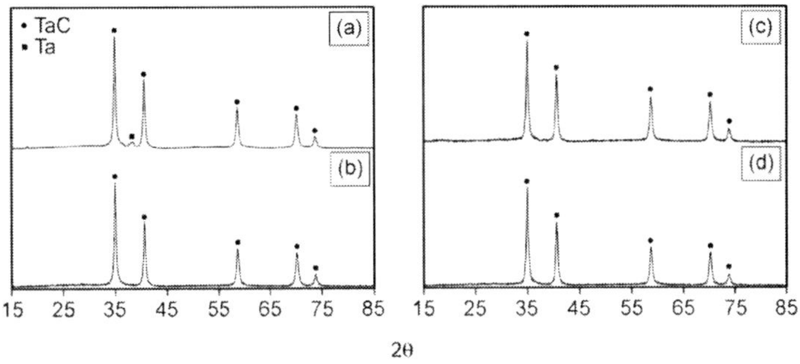

Figure 1. Comparison of X-ray diffraction patterns for (a) as-synthesized 50 g batch, (b) as synthesized 3 g batch, (c) 773 K heat-treated 50 g batch, and (d) 773 K heat-treated 3 g batch.

The particle size distributions for the 3 g batch and the 50 g batch are shown in Figure 2. The average particle size for the 3 g batch was 97 nm, while the average particle size for the 50 g batch was 73 nm. The ratio of average crystallite size to average particle size was reduced from 3.7 to 2.8, indicating lower levels of agglomeration and aggregation when scaling up. Scale-up showed no adverse effects on phase development and it did not cause grain growth or increased agglomeration or aggregation.

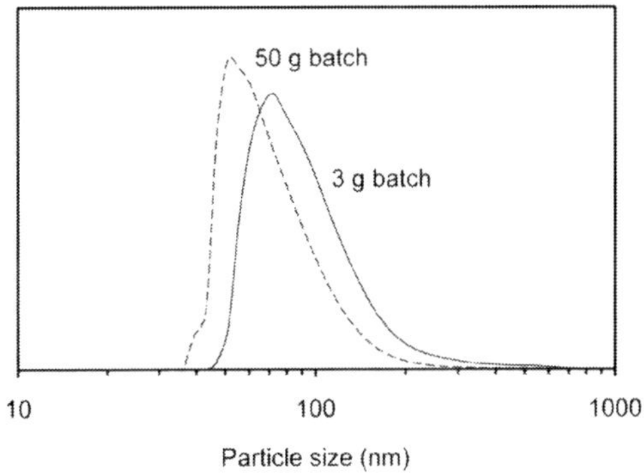

Figure 2. Particle size distributions for the 3 g and 50 g batches of TaC.

XPS spectra of the 3 g and 50 g batch powders are shown in Figure 3. There is little difference in the Ta 4f band, except that the bands were more intense and shifted to slightly lower binding energies when the synthesis was scaled up. For the 3 g batch, there is a Si 2p band that was observed, indicating contamination from the fused quartz tubes that were used. This peak is not present in the scaled-up batch because a stainless steel reaction vessel was used. Iron could be present in the samples because of the stainless steel container. More definitive evaluation of iron content and removal techniques (if necessary) could be useful. When scaling up, there is a significant increase in the C 1s band area with the appearance of an additional band centered near 288 eV. The increase in the carbon signal is likely related to the increase in tantalum signal and the two are a result of reduced impurities. The O 1s signal has two bands, one centered at approximately 530 eV and the other near 532 eV. In the scaled-up batch, the band at 532 eV was eliminated and is likely a result of transitioning to a stainless steel reaction vessel. The other band centered near 530 eV is not eliminated and could be caused by tantalum oxide, iron oxide, or adsorbed molecules on the surface.

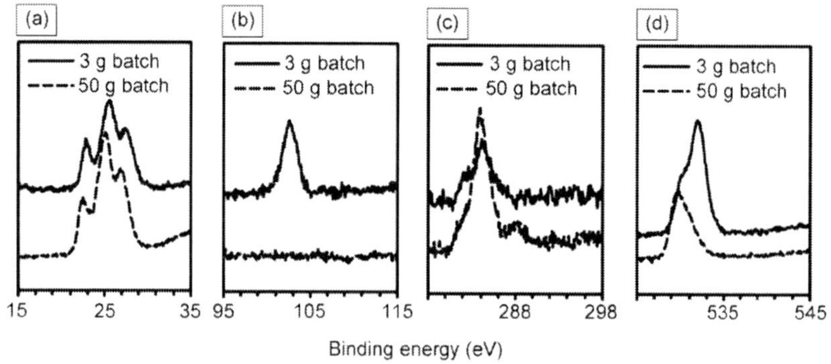

Figure 3. XPS spectra for (a) Ta 4f, (b) Si 2p, (c) C 1s, and (d) O 1s emissions.

TGA of the 3 g batch treated at different temperatures and comparing the effects of scaling are shown in Figure 4. In Figure 4a, it can be seen that the higher the heat treatment temperature, the less adsorbed species on the powders and that heat treatments of at least 600°C were necessary to completely eliminate weight losses. When comparing the two batch sizes in Figure 4b, the 3 g batch has more contaminant in the as-synthesized state than the 50 g batch. The 50 g batch began gaining weight at a lower temperature than the 3 g batch in both the as-synthesized and heat treated state. Results suggest concurrent oxidation and removal of adsorbed species.

The oxygen analysis showed that the oxygen content for the 50 g batch powders was 7.9% in the as-synthesized state. This was reduced to 6.1% on powders heat treated at 873 K for 3 hours in an argon atmosphere. This is a high content for non-oxide powders and can be detrimental to properties.

Figure 5 displays results for the TGA of the 50 g batch of TaC in a helium atmosphere with mass spectrometry. The figure shows that water is the most abundant species on the surface, but oxygen, carbon monoxide, carbon dioxide, and nitrogen are also burned off. The weight loss continues until about 500°C, adding to our previous assessment, oxidation and water removal can occur simultaneously over the temperature range of ~200-500°C. Removing the water from the surface by washing with oxygen-deficient water-displacing solvents prior to heat treatments is a potential method for reducing the oxygen content in the powders.

Lattice oxygen or oxide impurities on the surfaces may also be present as a result of the synthesis process. Therefore it is important to consider how to remove these sources for reducing the oxygen content. The traditional method of removing oxygen impurities from the surface involves leaching. Potential techniques for reducing oxygen content include using sulfuric acid, hydrofluoric acid, hydrochloric acid, alkaline hydroxides, or combinations of these at room temperature or slightly elevated temperatures. Use of acids also has the added advantage of allowing the preparation of special powder morphologies, such as cubic morphologies, which can result from etching of the powder surfaces. We are proceeding with these experiments and will present such results in a future communication.

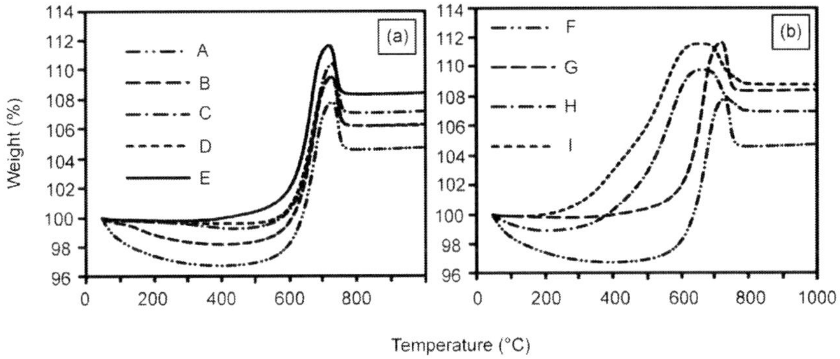

Figure 4. TGA results, performed in air, for (a) the 3 g batch subjected to various heat treatments and (b) a comparison between the 3 g batch and the 50 g batch in both the as-synthesized and heat-treated states. The treatments prior to TGA are provided in Table I.

TABLE I. HEAT TREATMENT SCHEDULES FOR FIGURE 4.

	Heat treatment		Heat treatment		Heat treatment
A	None	**D**	400°C, Air, 6 h	**G**	600°C, Argon, 6 h (3 g)
B	90°C, Vacuum, 6 h	**E**	600°C, Argon, 6 h	**H**	None (50 g)
C	225°C, Vacuum, 6 h	**F**	None (3 g)	**I**	600°C, Argon, 6 h (50 g)

Sintering of powders from the 3 g batches resulted in a reaction between the powders and the graphite SPS die and plungers. Furthermore, the resulting specimen separated into a dense metallic layer and a low-density powder layer. The deteriorated graphite and the two sides of the compact are shown in Figure 6. The reaction between the die and the plungers is a result of the silica contamination from the fused-quartz test tubes, which melted and then reacted during sintering. Fused-quartz reaction vessels are not suitable for this process and should be avoided unless procedures are implemented to remove silica from the powders.

The phenomenon observed in Figure 6 for the sintering of powders synthesized in the fused-quartz test tubes was not observed when sintering powders synthesized in the stainless steel beakers. Figure 7 displays the experimental temperature profiles and the plunger displacement rate curves that were observed during sintering of two 6 g compacts that utilized the powders synthesized in stainless steel beakers. The plunger displacement rate is directly related to the sintering rate of the compacts. One sample was heated to 1725°C (sample denoted SPS1) and the other to 1525°C (sample denoted SPS2). The displacement rate curves for the two compacts are similar and show that sintering of both samples stopped by ~1500°C and did not continue with higher temperatures, up to 1725°C.

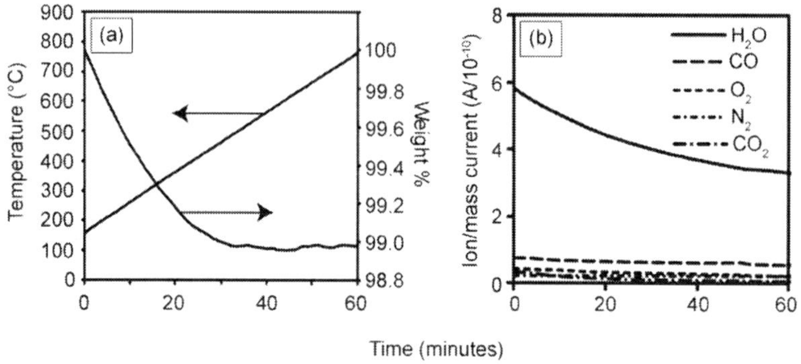

Figure 5. (a) TGA experiment of the 50 g batch in a helium atmosphere and (b) the corresponding mass spectroscopy.

Figure 6. Images showing the die, plunger, and two sides of the compact after SPS of powders synthesized in a fused-quartz test tube.

The lack of any displacement after 1500°C leads to an initial conclusion that the samples have become fully sintered. However, porosity is observed upon evaluating the microstructure of the first compact (SPS1), as is shown in Figure 8. There are large pores (several tens of microns) between large microstructural features (up to several hundred microns) that are shown by Figure 8 (a1 and b2).

The non-homogenous microstructure is a result of non-homogenous agglomeration of the powders in the as synthesized state. Sintering occurs within agglomerates more rapidly than amongst agglomerates due to the higher number of particle-particle contacts within the agglomerate. This causes non-homogenous sintering and therefore non-homogenous shrinkage to occur.

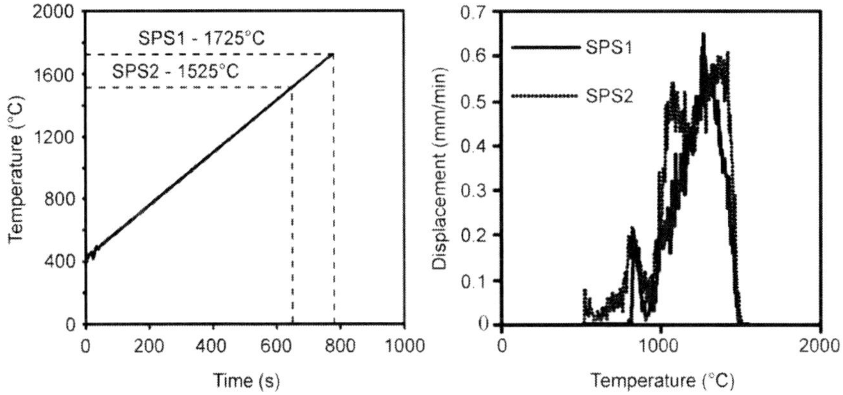

Figure 7. The measured temperature during SPS of tantalum carbide samples (SPS1 and SPS2) before cutting the power and pressure to the SPS unit. The corresponding displacement rate that was observed as the temperature was increased is also included.

The large microstructural features of higher density in Figure 8 (a1 and b1) are representative of the powder agglomerates. The agglomerates are of different shape; some are elongated while others are more equiaxed and most of them are irregular in shape. The non-homogeneity agrees with our previous assessment of the cause for the non-homogenous microstructure. As these large features sinter more rapidly than amongst each other, they shrink away from each other and produce the large voids between the agglomerates. Once formed, such large porosity is difficult to eliminate from a microstructure.

Further post-synthesis powder processing can be implemented to improve the microstructural homogeneity by achieving homogeneous agglomeration in the dry powder state. Spray drying, to granulate the powders, is the common method used to achieve homogeneous agglomeration for dry powder pressing processes. A simpler rudimentary approach is dry mixing of the powders in a mill to achieve a relatively narrow distribution of powder agglomerates that are spherical in nature. Homogeneous spherical agglomerates will deform properly in a dry pressing process to minimize the intra agglomerate porosity between the agglomerates. This, in combination with particle homogeneity within the agglomerate will ensure homogeneous shrinkage during sintering and lead to a homogenous microstructure.

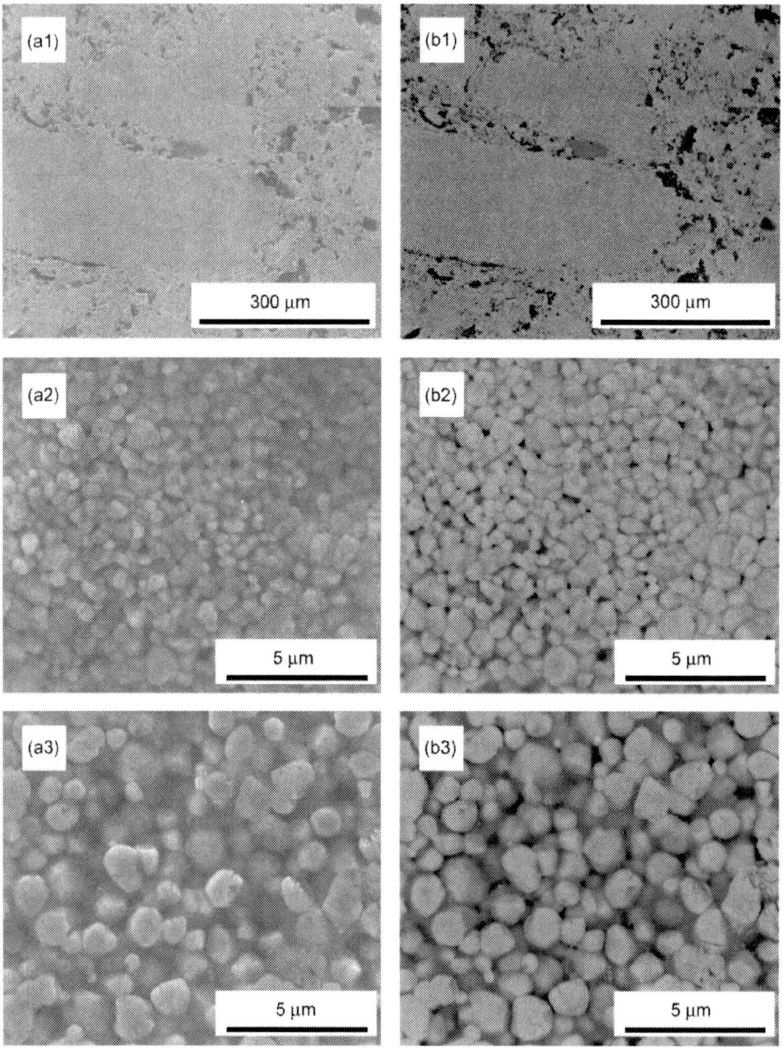

Figure 8. Various scanning electron micrographs taken in (a) secondary electron imaging mode and (b) the corresponding back-scattered electron images.

The remainder of the images in Figure 8 show the microstructural agglomerates taken at an initial magnification of 10,000x and highlight a few additional features of the microstructure.

Namely, the microstructure also contains sub-micron sized porosity, secondary phase of low density that is observed in back-scattered electron imaging mode, and inhomogeneous grain size distributions within the agglomerates.

Despite the higher density within the agglomerates, sintering within them is not complete. Sub-micron porosity is observed between the particles as is shown in Figure 8 (a2 and b2). Furthermore, many of the grains maintain a near spherical shape rather than a faceted grain structure typical of a fully dense microstructure. Sintering is far from complete even at a temperature of 1725°C and higher sintering temperatures or pressures may be necessary.

A secondary phase of low density is observed in some regions. This is demonstrated by Figure 8 (a3 and b3).

Grain size heterogeneity is also observed in the microstructure. This is best demonstrated by comparing image (a2) and (a3) or their respective back-scattered electron images in Figure 8. In (a2) or (b2), there are some grains as large as a micron, but a large number of the grains are submicron and as small as the starting particle size, whereas the majority of the grains in (a3) or (b3) are closer to 1-2 μm in size, which is an order of magnitude larger than the starting particle size. Images (a3) and (b3) are also the images where the low density phase is observed and this phase might facilitate grain growth. Starting particle heterogeneity, the low density phase, and the inherent nature of SPS can all be responsible for the grain size heterogeneity and will need further evaluation.

Figure 9 shows back-scattered images for sample SPS2. Similar microstructural features were observed for this sample, including spherical grains, a secondary phase, incomplete sintering, and a heterogeneous microstructure (pore and grain sizes). Comparing Figure 8 and Figure 9 we can see that a smaller grain size was achieved for SPS2, which might be expected from the lower sintering temperature. The largest grains are near a micrometer, but the majority of the grains are submicron and closer to the starting particle size. The sintered regions seem denser and the secondary phase (darker gray regions) is more apparent for SPS2. The low density of SPS1 and less obvious characteristic of the secondary phase might indicate that the secondary phase is being at least partially removed or reacting at higher sintering temperatures.

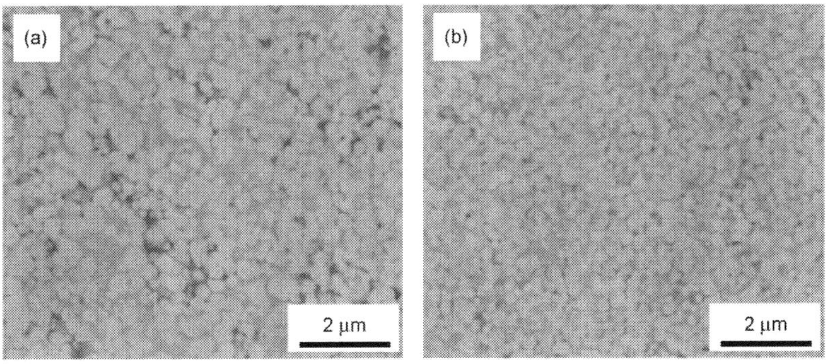

Figure 9. Various scanning electron images of sample SPS2 taken in back-scattered electron imaging mode showing (a) region with larger grains and (b) a region with smaller grains.

The X-ray diffraction pattern for sample SPS1 is shown in Figure 10. Only tantalum carbide was observed in the pattern. The secondary phase might not be observed if it were amorphous or if the elements of the phase do not provide significant X-ray scattering, such as if residual lithium from the synthesis process remains. Thus, the sample was oxidized at 1000°C to try to crystallize out the secondary phase that was observed with SEM. The X-ray diffraction for the resulting powder after oxidation is also shown in Figure 10. In addition to tantalum carbide and oxide phases, lithium tantalate was also observed, indicating that residual lithium is left in the powders. The lithium containing phase may also be a contributing factor to the high oxygen content that was observed for these powders. The powder washing procedures need to be revisited and optimized. Our future communications will address this issue.

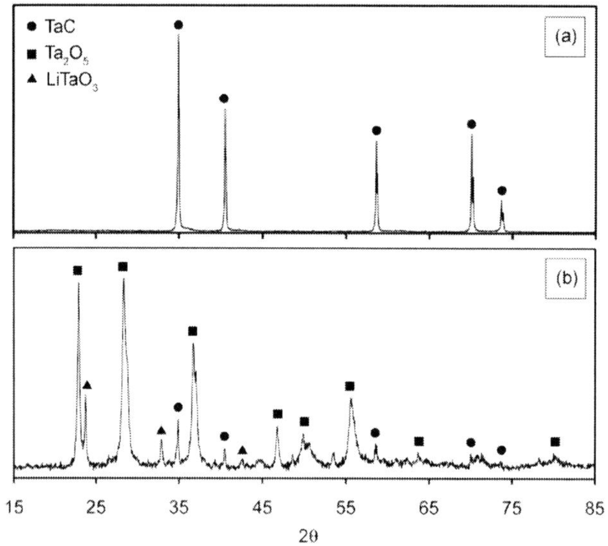

Figure 10. X-ray diffraction pattern for sample SPS1 (a) before oxidation and (b) after oxidation.

CONCLUSIONS

It has been shown that it is possible to scale-up the solvothermal synthesis process while maintaining nano-sized powders with no detrimental impact on the powder's characteristics. Phase purity, average crystallite size, and compound stoichiometry were all maintained during scale-up. Particle size distribution and surface contaminants were both decreased. Oxygen content of the powders is high and needs to be further investigated in order to decrease it.

Sintering trials revealed the need to optimize the washing procedure before carrying out further sintering experiments. Optimization of the washing procedures can eliminate lithium from the powders, reduce the oxygen content, and may also lead to alternate morphologies via leaching and etching. Furthermore, forming homogenous powder agglomerates is necessary to form a more homogenous microstructure.

ACKNOWLEDGMENT

This work was supported by a grant from the National Science Foundation under grant numbers CMMI 0645225 and CMMI 0913373. The assistance of S.D.D.L. Torre with some of the spark plasma sintering work is gratefully acknowledged.

REFERENCES

[1]. D. Sciti, L. Silvestroni, S. Guicciardi, D. Dalle Fabbiche, and A. Bellosi, *J. Mater. Res.*, **24** [6] 2056-65 (2009).

[2]. F. Peng, L Han, H. Fu, and X. Cheng, *Phys. Status Solidi B*, **246** [7] 1590-6 (2009).

[3]. B.-R. Kim, K.-D. Woo, J.-M. Doh, J.-K. Yon, and I.-J. Shon, *Ceram. Int.*, **35** [8] 3395-400 (2009).

[4]. Y. Zhang, C. Zhang, H. Hu, and Y. Zhou, *Key Eng. Mater.*, **362-372** [2] 1771-3 (2008).

[5]. X. Zhang, G.E. Hilmas, and W.G. Fahrenholtz, *J. Am. Ceram. Soc.*, **91** [12] 4129-32 (2008).

[6]. J. Noffsinger, F. Giustino, S.G. Louie, and M.L. Cohen, *Phys. Rev. B.*, **77** [18] 180507/1-/4 (2008).

[7]. A.A. Lavrentyev. B.V. Gabrelian, V.B. Vorzhev, I.Y. Nikiforov, O.Y. Khyzhun, and J.J Rehr, *J. Alloys Compd.*, **462** [1-2] 4-10 (2008).

[8]. X. Guo, L. Li, Z. Liu, D. Yu, J. He, R. Liu, B. Xu, Y. Tian, and H.-T. Wang, *J. Appl. Phys.*, **104** [2] 023503 (2008).

[9]. J.P. Kelly, R. Kanakala, and O.A. Graeve, *J. Am. Ceram. Soc.*, **93** [10] 3035-8 (2010).

[10]. J.P. Kelly and O.A. Graeve, *J. Am. Ceram. Soc.*, **94** [6] 1706-15 (2011).

[11]. A.L. Bowman, *J. Phys. Chem.*, **65** [9] 1596-8 (1961).

Mater. Res. Soc. Symp. Proc. Vol. 1373 © 2012 Materials Research Society
DOI: 10.1557/opl.2012.290

Thermal Behavior of Celsian Ceramics Synthesized from Coal Fly Ash

Jorge López-Cuevas[1], David Long-González[2], Carlos A. Gutiérrez-Chavarría[1]
[1]Cinvestav-Saltillo, Carr. Saltillo-Monterrey Km 13.5, Ramos Arizpe, Coahuila, México, 25900
[2]RHI REFMEX, S.A. de C.V., Carr. Saltillo-Monterrey Km. 9, Coahuila, México, 25900

ABSTRACT

Celsian with a chemical composition of $Ba_{0.75}Sr_{0.25}Al_2Si_2O_8$, is synthesized by using coal fly ash (byproduct of a Mexican coal-burning power plant, composed mainly by SiO_2 and Al_2O_3) as main raw material. The thermal behavior of the synthesized material is evaluated by differential (DTA) and gravimetric (TGA) thermal analyses as well as by heating microscopy; its coefficient of linear thermal expansion (CTE) is also determined. Heating microscopy shows that cylinders of compacted powdered Celsian start sintering at ~1140 °C, which is associated with a considerable contraction occurring up to 1500 °C. The mean CTE value of the material in the temperature range of 30-1100°C is slightly affected by the synthesis conditions employed. Synthesis at 1400 or 1300 °C during 10 h, with a pre-calcination step at 900 °C/5h in both cases, produce mean CTE values of 5.15×10^{-6} and 5.43×10^{-6} °C^{-1}, respectively. On the other hand, Celsian synthesized at 1400 °C/10 h, without the pre-calcination step, has a mean CTE value of 5.25×10^{-6} °C^{-1}. Lastly, the DTA/TGA analysis of the synthesized material shows that a slight weight gain takes place from room temperature to 1100°C, which is followed by a slight weight loss up to 1300°C. This is attributed to oxidation and evaporation of some of the impurities present in the material.

INTRODUCTION

Barium-strontium aluminosilicate solid solutions ($Ba_{1-x}Sr_xAl_2Si_2O_8$) are interesting materials for structural applications at high temperatures, especially in their monoclinic polymorphic form known either as Celsian or as Monocelsian. These materials exhibit high refractoriness, low thermal expansion coefficient, good resistance to oxidation and to slag attack, as well as good thermal shock resistance [1,2]. These solid solutions are formed when BaO is partially substituted by SrO in the crystalline structure of barium aluminosilicate ($BaAl_2Si_2O_8$, BAS); this process constitutes an effective way to promote the formation of the monoclinic BAS form corresponding to the so-called Celsian or Monocelsian phase [3]. Without the addition of SrO, or of other mineralizing agents that cause a similar effect [4], a hexagonal polymorphic form of BAS (Hexacelsian) is very likely to be formed instead of Celsian, and it frequently remains in a metastable state at temperatures at which the latter is the equilibrium phase (from room temperature to 1590°C) [5]. Hexacelsian is undesirable due to its relatively high thermal expansion coefficient and because it undergoes a conversion into an orthorhombic form at ~300°C, which is associated with a volume change of ~3-4%. This causes microcracking in the material, negatively affecting its final mechanical properties [2,6,7].

In this work, Celsian with a chemical composition of $Ba_{0.75}Sr_{0.25}Al_2Si_2O_8$ is synthesized by using coal fly ash (byproduct of a Mexican coal-burning power plant, composed mainly by SiO_2 and Al_2O_3) as main raw material. The thermal behavior of the synthesized materials is evaluated in terms of its weight and dimensional stability.

EXPERIMENTAL DETAILS

The raw materials employed in this study are (1) Coal Fly Ash (CFA) from the "José López Portillo" power plant (Nava, Coahuila, México). This material is ground in a ball mill during 5h and subjected to a wet magnetic separation process to eliminate as much iron oxide as possible, producing a mean particle size of ~5 μm. It contained quartz, mullite and glass as main phases, and its chemical composition is (in wt.%): 64.59% SiO_2, 25.73% Al_2O_3, 2.57% Fe_2O_3, 2.12% CaO, 0.59% TiO_2, 0.48% MgO, 0.46% Na_2O, 0.37% K_2O, and 1.11% of other oxides (SrO, ZrO_2, CuO, NiO, SO_3 and P_2O_5). Ignition of this material caused a weight loss of 1.94% due to elimination of free C. (2) $BaCO_3$, with a purity of 99.43 wt.% and mean particle size of ~3 μm (Alkem, México).(3) $SrCO_3$, with a purity of 97.86 wt.% and mean particle size of ~4 μm (Solvay, México). (4) Al_2O_3, with a purity of 99.99 wt.% and mean particle size < 1 μm (HPA-0.5 Sasol, USA).

Celsian of the desired composition is prepared using a suitable mixture of raw materials followed by cold isostatic pressing at 200 MPa to produce prismatic monoliths with approximate dimensions of 70x30x13 mm. Then, the monoliths are subjected to a solid state reaction during 5 or 10h at temperatures of 1300 or 1400°C. Some of the samples are subjected to a calcination step at 900°C/5h in order to decompose the carbonates contained in the initial mixtures. Phase analysis of the sintered samples is carried out by X-Ray Diffraction (XRD) using a Siemens D-5000 XRD apparatus and employing CuKα radiation in the 2θ range of 10-70°, with a step size of 0.05° and a counting interval of 3s. This analysis corroborated the formation of the desired phase in the sintered samples (Figure 1).

The thermal behavior of the synthesized materials is evaluated by differential (DTA) and gravimetric (TGA) thermal analyses as well as by heating microscopy. Linear thermal expansion coefficients (CTE) are also determined. The thermal analyses are carried out up to a maximum temperature of 1400°C employing a PERKIN ELMER DTA 7 apparatus. Platinum crucibles are used and the experiments are performed at a heating rate of 10°C/min under an atmosphere of still ambient air. The CTE is determined in the temperature range from room temperature to 1200°C using a Netzch DIL 402 PC dilatometer, employing an alumina rod as standard and sample bars with dimensions of 40x5x5 mm. The heating microscopy studies are carried out by using an EM201 Hesse Instruments heating microscope fitted with a computerized image analysis system and an electric furnace 1750/15 (Leica, Germany). This technique allows following the sintering process of compacted powdered synthesized Celsian in the temperature range of 25-1500°C at a heating rate of 5°C/min.

During a run on the heating microscope, images of the sample silhouette are projected through a transparent quartz window and automatically recorded by the image analysis system. Then, the system's software analyzed, from the recorded images, any geometrical changes underwent by the sample during heating, such as variations in height, width and projected area. For this study, the synthesized samples are first ground and then cylindrical specimens with diameter and height of 1.5mm are obtained by gentle uniaxial pressing. To carry out the tests, the compacted cylinders are placed on top of a ceramic plate and introduced into the heating microscope. Temperature is measured in the proximity of the specimens with a Pt3%Rh–Pt/Pt10%Rh thermocouple making contact with the supporting ceramic plate.

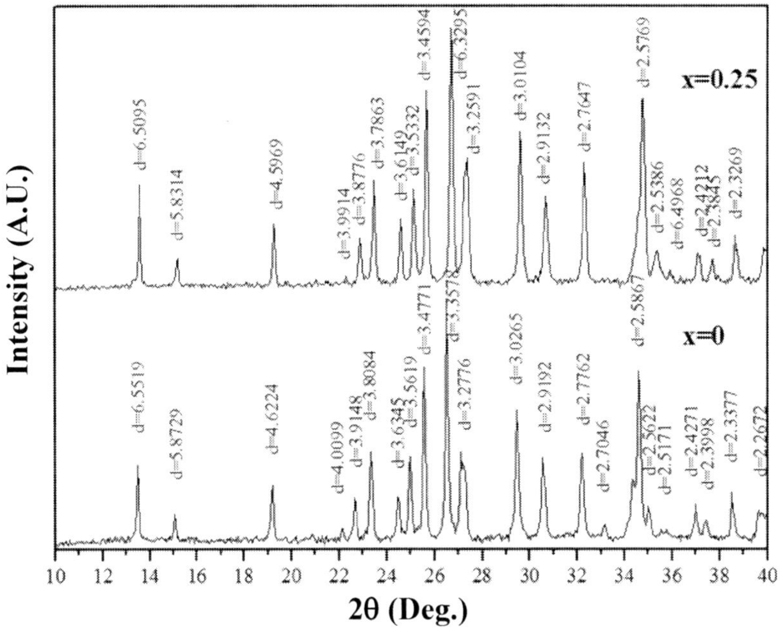

Figure 1. XRD patterns of Celsian monoliths with X = 0 (for comparison only) and X = 0.25 (synthesized material), pre-calcined at 900°C/5h and sintered at 1400°C/10h.

RESULTS AND DISCUSSION

Thermal Analysis

Figure 2 shows that Celsian synthesized at 1400°C/5h and with a pre-calcination step at 900°C/5h had a weight gain of ~2% from room temperature to 1100°C, with a subsequent weight loss of ~1.5% occurring up to 1300°C. This is likely due to oxidation and evaporation of some of the impurities present in the synthesized material.

Heating Microscopy

The heating microscopy images presented in Figures 3 and 4 show that the compacted powdered Celsian material started to sinter at ~1140°C and that this process is accompanied by a contraction of ~24.5% of the sample cylinder up to the final temperature of the test (1500°C). Full melting of the material is not detected up to the maximum temperature, but partial melting can not be completely ruled out due to a likely fluxing effect from some of the impurities present in it.

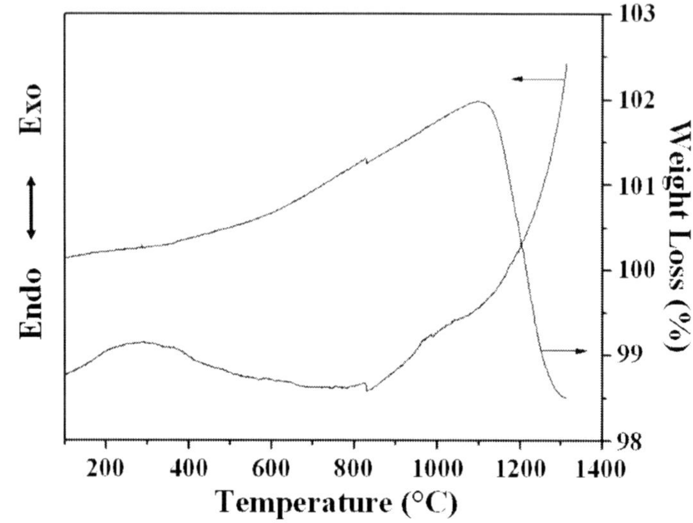

Figure 2. DTA and TGA curves obtained for Celsian sintered at 1400°C/5h, pre-calcined at 900°C/5h.

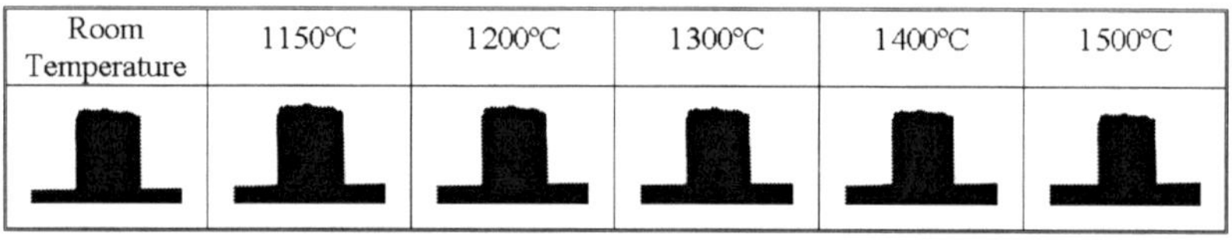

Figure 3. Photographs of sample silhouettes taken at different temperatures for Celsian synthesized at 1400°C/5h, pre-calcined at 900°C/5h.

Coefficient of Thermal Expansion (CTE)

As can be seen in Figure 5 and Table 1, the mean CTE value of the synthesized material in the temperature range of 30-1100°C is slightly affected by the synthesis conditions employed. Synthesis at 1400°C or 1300°C during 10h, with a pre-calcination step at 900°C/5h in both cases, produced mean CTE values of 5.15 x 10^{-6} and 5.43 x 10^{-6} °C^{-1}, respectively. In contrast, Celsian synthesized at 1400°C/10h, without pre-calcination, had a mean CTE value of 5.25 x 10^{-6} °C^{-1}.

In general, the highest mean CTE values are obtained for the materials sintered at the lowest temperature, which is attributed to a lack of densification in this case.

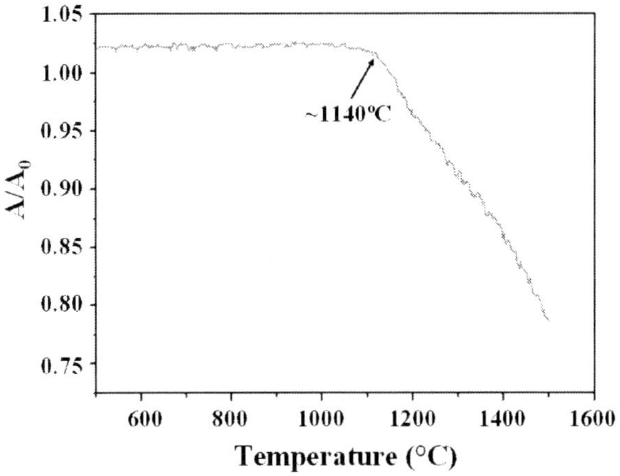

Figure 4. Variation in the projected area of a cylinder of compacted Celsian powder photographed in Figure 3.

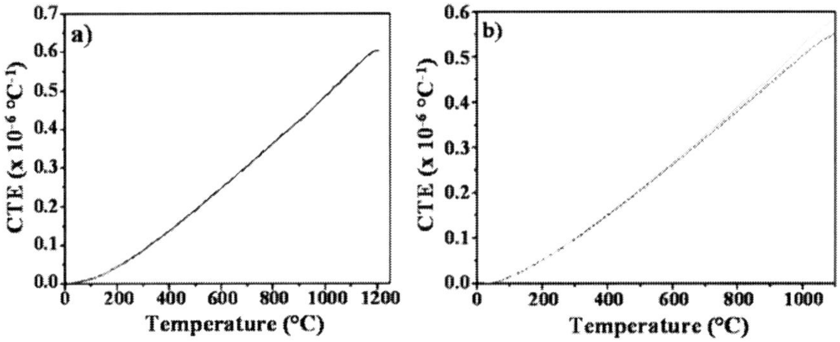

Figure 5. CTE values obtained for: a) Celsian sintered at 1400°C/10h in a single step, and b) Celsian sintered at 1300°C/10h (dotted line) or at 1400°C/10h (full line), in both cases with pre-calcination at 900°C/5h.

CONCLUSIONS

The weight of the synthesized Celsian is relatively stable at high temperatures, except for a slight oxidation followed by evaporation of a small amount of some of the impurities present in the material. Heating microscopy showed that the synthesized compacted powdered Celsian started to sinter at ~1140°C, with a considerable associated contraction occurring up to the final temperature of the test (1500°C). The mean CTE value of the materials, in the temperature range of 30-1100°C, is only slightly affected by the synthesis conditions employed. In general, the highest mean CTE values are obtained for the materials sintered at the lowest temperature employed, which is attributed to a lack of densification in this case.

Table 1. Mean CTE values (x 10^{-6} $°C^{-1}$) obtained for Celsian materials synthesized under different processing conditions.

Conditions of thermal treatment	30-900°C	30-1000°C	30-1100°C
1400°C/10h, single step	4.82	4.96	5.12
1300°C/10h, with pre-calcination at 900°C/5h	5.21	5.40	5.43
1400°C/10h, with pre-calcination at 900°C/5h	5.09	5.18	5.15

ACKNOWLEDGMENTS

The authors express their gratitude to CONACYT and Cinvestav-Saltillo for the financial support and experimental facilities provided for the development of this work.

REFERENCES

1. X.-D. Zhang, K.H. Sandhage and H.L. Fraser, *J. Am. Ceram. Soc.* **81**, 2983-2997 (1998).
2. J.S. Moya Corral and G. Verduch, *Brit. Ceram. Trans. J.* **77**, 40-44 (1978).
3. N.P. Bansal, M.J. Hyatt and C.H. Drummond III, *Ceram. Eng. Sci. Proc.* **12**, 1222-1234 (1991).
4. K.T. Lee and P.B. Aswath, *Mater. Sci. Eng. A* **352**, 1-7 (2003).
5. H.C. Lin and W.R. Foster, *Am. Mineral.* **53**, 134-144 (1968).
6. B. Yoshiki and K. Matsumoto, *J. Am. Ceram. Soc.* **80**, 2021-2029 (1997).
7. N. Frety, A. Taylor and M.H. Lewis, *J. Non-Cryst. Solids* **195**, 28-37 (1996).

Mater. Res. Soc. Symp. Proc. Vol. 1373 © 2012 Materials Research Society
DOI: 10.1557/opl.2012.291

Surface Study of Vitreous Carbon Obtained from Glassy Carbon Powder

F. Dondeo[1], A. J. Damiao[1], W. Miyakawa[1], F. Nascimento[1], L. Fernandes[1], S. Oishi[2], E. Botelho[2]
[1]Photonics Division/Institute for Advanced Studies/DCTA, Trevo Coronel Aviador José Alberto Albano do Amarante, 1, São José dos Campos, CEP 12228-001, Brazil.
[2]Department of Materials and Technology, São Paulo State University, Av. Ariberto Pereira da Cunha, 333, Guaratinguetá-SP, CEP 12516-410, Brazil; e-mail: dondeo@ieav.cta.br

ABSTRACT

This work presents a surface study of monolithic vitreous (or glassy) carbon - MVC - obtained from vitreous carbon powder. Defective MVC pieces are crushed in a ball mill and size classified by sifting. The MVC powder is mixed with furfuryl-alcohol resin and compacted in a mould using a hydraulic press. Samples with different powder granulometries are produced in this way and carbonized in a furnace under nitrogen atmosphere. Complete carbonization of the "powder" is achieved in only one day and losses due to breakage of the pieces is less than 5%. These results compare very favorably with respect to traditional MVC production methods where full carbonization may require up to seven days and losses due to breakage can be as high as 70%. After carbonization, samples are sanded and polished. Surface roughness and microstructure are characterized by light microscopy. Porosity is quantified from micrographs using *ImageJ* software and nanometric height variations are measured by atomic force microscopy.

INTRODUCTION

Vitreous carbon (or glassy carbon) is a carbonaceous material with high stiffness and low density suitable for applications in the aerospace industry. In addition, this material exhibits good electrical conductivity, good resistance to high temperatures in non-oxidizing atmospheres and biocompatibility [1]. Two main forms of vitreous carbon are available: monolithic (MVC) and reticulated (RVC).

RVC has extremely high porosity (~95%) and is suitable for electrode manufacturing applications due to the large internal surface area and the electrical conductivity exhibited by this material. Vitreous carbon can also be used in the form of powder for various electrochemical [3, 4] and tribological [5] applications and in the manufacture of conductive glues [6] and carbon activated materials [7].

MVC is a compact material and porosity is usually unacceptable. The main limitation of MVC is that products thicker than 10 mm are difficult to prepare [2]. The traditional method to produce MVC consists in the polymerization of a carbon rich resin, e.g. furfuryl alcohol, phenolic or furfuryl-phenolic resin, followed by slow carbonization in a non-oxidant atmosphere (usually nitrogen gas or vacuum), at temperatures above 1000ºC. However, in this process, irreversible mechanical damage is usually observed for sample thickness exceeding about 10 mm.

In this work, a new processing route for the production of monolithic vitreous carbon is developed with the aim of overcoming some of the difficulties usually found in traditional methods involving high temperature polymerization of C-rich resins, e.g. low yields, slow production rates and thickness limitations. Both qualitative and quantitative surface studies of the produced MVC samples from glassy carbon particles are presented and discussed. To the knowledge of the authors, this is the first time that the surface microstructure characteristics of MVC are reported.

EXPERIMENT

Samples of MVC are produced by the traditional method of carbonization of cured furfuryl alcohol resin. Carbonization is carried out in a computer controlled tubular furnace under nitrogen atmosphere. The maximum temperature of the carbonization cycle ranged from 1050 to 1150 °C. The complete cycle lasted one whole week. As the majority of the samples (nearly 70%) are broken during this process due to the large amount of porosity, the damaged MVC material is recycled to produce new MVC samples. Broken parts are then crushed in a ball mill, and the obtained glassy carbon powder is classified by size in a vibrating sieve stack. By using six sieves with aperture sizes of 149, 105, 74, 62, 44 and 37 μm, respectively, seven groups of glassy carbon powder are obtained: one group with particles smaller than 37 μm, another with particles between 37 and 44 μm, and so on, up to the group with particles larger than 149 μm.

Furfuryl alcohol resin is mixed with 2 wt% of p-toluenesulfonic acid catalyst and centrifuged at 3000 rpm during 7 minutes to reduce the presence of bubbles (large pores) in the resin. For each group of powders, one third (in mass) of centrifuged resin is mixed with the powder during 5 minutes. This powdery mixture is pressed in a hydraulic press applying 30 MPa of pressure during 2 minutes. After removal from the mould, disk samples, 33 mm in diameter and 6 mm thick, are polymerized at room temperature during two days. Next, the discs are carbonized in the same furnace used before. In this case, the overall carbonization process did not exceed two days. The heating rate is nearly five times higher, because majority of the mass is already carbonized in the first heating process.

The surfaces of the vitreous carbon samples obtained from this alternative process are machined and then polished using 0.05 μm alumina powder. The surface microstructure of the samples is characterized by light microscopy (Polyvar-Met 2 *Leica*), confocal scanning laser microscopy (Lext OLS3100, *Olympus*) and atomic force microscopy (SPM 9500J3, *Shimadzu*) operating in the contact mode. Porosity of the samples is measured on five images obtained at random locations on the sample surface. These images are analyzed using *ImageJ* software. Roughness is evaluated using a *Taylor Robson* PGI-1000 perfilometer. Five measurements with 2.5 mm of length cutoff are performed on each sample.

RESULTS AND DISCUSSION

One of the main results of this alternative method is the increase in efficiency for the production of MVC samples using glassy carbon powder. Nearly 95% of the produced MVC samples did not break and did not exhibit external cracks after carbonization. This is an important distinction with respect to the traditional processing route that presents low efficiency.

Another significant difference is the lack of thickness restriction: several samples 10 to 15 mm thick are successfully produced. Therefore, this method is very advantageous in terms of increasing efficiency and critical thickness.

Typical surface confocal scanning laser microscopy images of the microstructures of samples prepared from the finest (less than 37 μm) and the coarsest (above 149 μm) powders are presented in Figure 1. Although porosity is considerable in both cases, pores size is much smaller – more than one order of magnitude - in the sample produced from finer powders.

Figure 1. Confocal scanning laser microscopy images of polished surfaces of MVC samples; a) corresponds to the finest powders (diameter less than 37 μm) and b) corresponds to the coarsest powders (diameter larger than 149 μm).

Porosity values, measured on images obtained at a magnification of 25X, are shown in Figure 2 as a function of the grain size of the glassy carbon powder used to manufacture the MVC. The grain size of the glassy carbon powder seemed to have little effect on the measured surface porosity values. It may be noted in Figure 1 that the pore size decreases when finer powder is used to manufacture the MVC. However, the results of Figure 2 show that the number of pores increases in such a way that the apparent porosity remains nearly constant.

The value of porosity corresponding to MVC produced by the traditional method is only 0.98%, which is 28 times lower than that for the new method here presented. At present, for applications where low surface porosity is detrimental, such as for mirrors substrates, the traditional method is still more advantageous.

The effect of grain size of the glassy carbon powder used to manufacture the MVC on the roughness of the samples is shown in Figure 3a. Arithmetic mean roughness (Ra) values varied from 0.7 to 5.1 μm and exhibited a trend to increase with grain size. For comparison purposes, in the origin of the graph (zero mm), the typical value of surface roughness of samples obtained by the traditional method (0.06 μm) is shown. Therefore, the smallest roughness value obtained by using glassy powder to produce MVC is approximately 12 times larger than that obtained by the traditional method. Thus, in the same way as porosity, for specific applications are low surface roughness is demanded, the traditional method is preferable. Although not shown, the Rz and Rzmax parameters are also measured and exhibited the same dependence on grain size as the Ra parameter. Figure 3b shows a typical roughness profile from a MVC sample produced from glassy powders. Despite the apparent flatness of the surface, very deep pores, tens of micrometers in depth, contributed to the high values of roughness measured.

While porosity is related to the relative area occupied by pores, roughness depends on pore depth. The results show that the samples produced with larger grains exhibit deeper pores.

However, the relative area occupied by the pores does not depend on the grain size for the experimental conditions of this work.

Microstructural features measured on AFM images are show in Figure 4. Dark areas correspond to deeper regions of the surface. The sample produced from the finest glassy C powders (diameter less than 37 μm) exhibited small holes with diameters of about 2 μm (Figure 4a). These small features are also observed in Figure 1a.

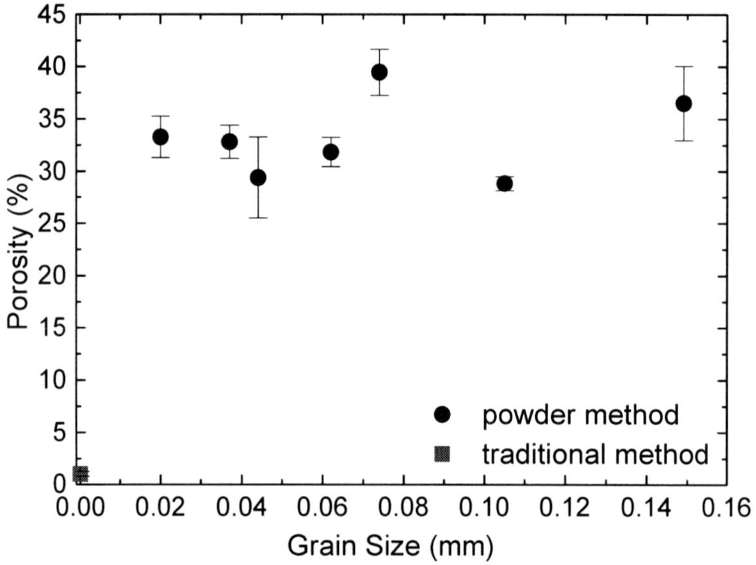

Figure 2. Effect of powder grain size on the apparent surface porosity of MVC samples produced by the "powder" method. The magnification employed to measure porosity is 25X. The square dot corresponds to a sample prepared by the traditional method.

As shown in Figure 4b for the case of the sample produced from the coarsest glassy C powders (diameter greater than 149 μm), an AFM image of a region of the same size showed only a section of a large hole. While height variations in the left image are of the order of hundreds of nanometers, in the right image this value reached the micrometer range. These results show clearly that porosity of samples produced from coarse glassy C powders is also larger when examined at high magnification.

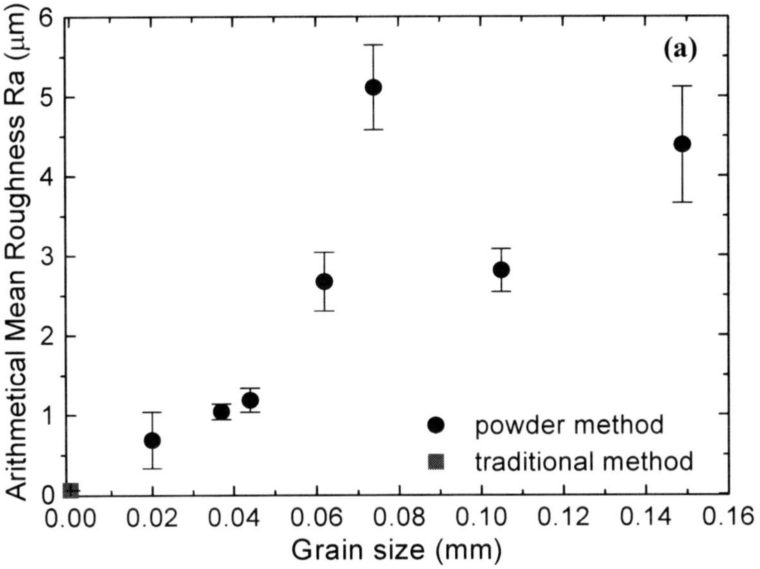

b)

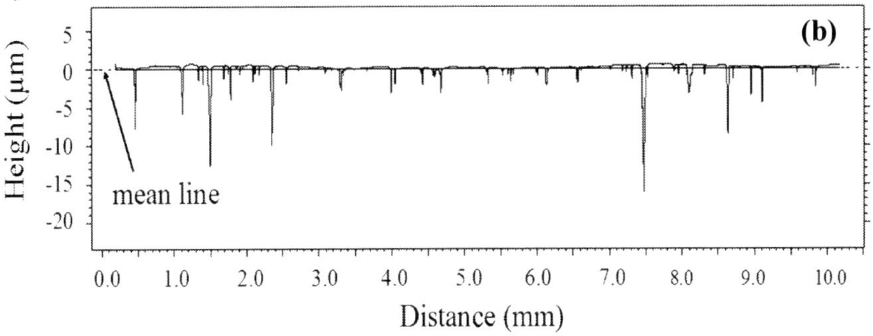

Figure 3. a) Effect of grain size of the glassy powder used to manufacture the MVC on the mean arithmetic roughness (Ra). The square dot corresponds to a value typically observed in samples of MCV produced by the traditional method; b) typical roughness profile showing a flat pattern with deep valleys that contributes to the high roughness values.

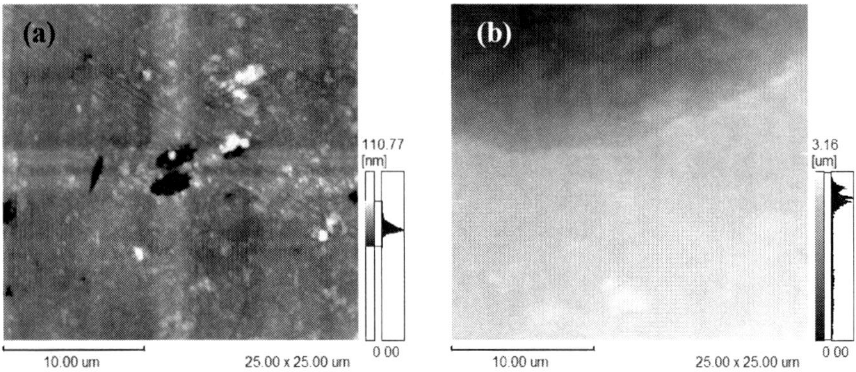

Figure 4. Atomic force microscopy images of polished surfaces of MVC samples produced from: (a) fine glassy C particles (diameter less than 37 μm) and (b) coarse glassy C particles (diameter greater than 149 μm).

CONCLUSIONS

Defective MVC samples can be successfully recycled to produce new MVC samples with the advantage of requiring shorter processing times (two days) than the traditional method (one week). The pieces produced did not exhibit macroscopic mechanical damage and, apparently, they can be produced with no thickness restriction. While porosity has not changed significantly with the powder particle size, roughness increased with the particle size and smoother surfaces can be produced using smaller grains. However, both porosity and roughness in MVC produced by the traditional method are still significantly smaller.

ACKNOWLEDGMENTS

The authors wish to thank to FINEP Project 555705/2009-9 and CNPq Projects 555705/2009-9 559925/2010-7, 384334/2009-1, 183867/2009-2, and 181958/2011-2 for the financial support.

REFERENCES

1. G. M. Jenkins and K. Kawamura, *Polymeric carbons – carbon fiber, glass and char*, Cambridge, New York (Cambridge University Press, 1976) pp. 52-160.
2. J. Zarzycki, "Glasses and Amorphous Materials", *Materials Science and Technology – a Comprehensive Treatment* **9**, ed. R. W. Cahn, P. Haasen and E. J. Kramer (Weinheim, New York, Basel, Cambridge, 1991) pp. 551-572.
3. M. Kurihara, S. Maruyama, K. Oh'e and T. Ishigaki, *Chemistry Letters* **8**, 715 (1998).
4. U. Anik, S. Cevik and M. Pumera, *Nanoscale Research Letters* **5**, 846 (2010).
5. J. Myalski and J Sleziona, *Journal of Materials Processing Technology* **175**, 291 (2006).
6. S. Gama and R M. Peçanha, patent BR200502173-A (May/2005).
7. A. Braun, M. Bartsch, B. Schnyder, R. Kotz, O. Haas and A. Wokaun, *Carbon* **40**, 375 (2002).

Mater. Res. Soc. Symp. Proc. Vol. 1373 © 2012 Materials Research Society
DOI: 10.1557/opl.2012.292

Development of Silicon Carbide Substrates for Aerospace Applications

Antonio J. Abdalla [1], Álvaro J. Damião[1], Elson Campos[2], Jerusa G.A.Santana[2], Marcelo C.Vicentini[2], Thiago A.Trevisan[2], Henrique Inoue[2], Marcos V.R. Santos[3], Francisco C.L.Melo[4]
[1]Institute of Advanced Studies - Photonics Division, Rodovida dos Tamoios, km 5,5 ,12.228-001 Sao Jose dos Campos – SP, Brazil
[2]Faculty of Engineering Guaratinguetá - FEG / UNESP, 12.500-000 - Guaratinguetá – SP, Brazil
[3]Aeronautical Institute of Technology – ITA - São José dos Campos – SP - Brazil
[4]Institute of Aeronautics and Space - Materials Division - São José dos Campos – SP - Brazil

ABSTRACT

Payload and high-tech are important characteristics when the goals are aerospace applications. The development of the technologies associated to these applications has interests that transcend national boundaries and are of strategic importance to the nations. Ultra lightweight mirrors, supports and structures for optical systems are important part of this subject. This paper reports the development of SiC substrates, obtained by pressing, to be applied on embedded precision reflective optics. Different SiC granulometries, having YAG as sintering additive, were processed by: ball milling, drying and deagglomeration, sift, uniaxial and isostatic pressing, and, finally, argon atmosphere sintering at 1900°C. Different porosities were obtained according to the amount of organic material added. Into one side of the samples pellets of organic material were introduced to generate voids to reduce the weight of samples as a whole. The substrates were grinding and polished, looking for a SiC surface having low porosity, as porosity is directly related to light scattering that should be avoided on optical surfaces. Laser surface treatments were applied (using or not SiC barbotine) as a method to improve the surface quality. The samples were characterized by optical and laser confocal microscopy, roughness measurements and mechanical tests. The results are very promissory for future applications.

INTRODUCTION

The use of precision optical components for aerospace and commercial uses requires important features such as: low density, excellent surface finish and rigidity. In this context, studies covering the development, fabrication and characterization of artifacts reflectors (mirrors) through the use of materials such as glassy carbon and silicon carbide.

In order to obtain ceramics with different porosities and for applications such as reflectors, samples are made of SiC ceramics. In this process, YAG is used YAG as the sintering additive. Features such as high hardness, high melting point, chemical stability at room temperature, high thermal stability and resistance to high temperatures (1, 4), are decisive for the choice of SiC. Then, introducing YAG additive, is motivated by factors such as reduction in sintering temperature and densification of the SiC samples, resulting in a ceramic body with improved mechanical strength (2).

All samples so produced by the technique of pressing, a fact which enabled the formation of samples with improved mechanical strength, surface finish with more regular and smaller pores, due to more efficient packing of the particles. However, as the use of a reflector system requires an excellent surface finish, the ceramics were subjected to a laser treatment with and without addition of a colloidal suspension (slurry) of silicon carbide.

Among other techniques of surface treatment, the use of a laser is an important tool. In this work, in order to improve the surface characteristics of ceramics, we use a low power CO_2 laser to melt the material surface and reduce the size and number of pores and, subsequently, with the same purpose, this is used for sintering of an additional layer of SiC deposited on the samples.

EXPERIMENT

The ceramic used in this study are made with two types of silicon carbide: β-SiC produced and processed by the *Raw Materials*, (designated SiC^1) β-SiC: GRADE BF-12 type, produced by *C. Hermann Starck* (HCST) (designated SiC^2). Some additives have been incorporated in the composition of the mixture to improve the properties of the final product: YAG (Yttrium Aluminium Garent) - 7.6% by mass, the PVal binder (polyvinyl alcohol), manufactured by *VETEC Química Fina do Brasil* and Bio deflocculant Denvercril 300 produced by *DENVER Resinas*. The choice of two types of SiC and the percentage of YAG is motivated by results previously obtained by the Division of Materials - AMR / DCTA and Marins (3).

The samples are processed by wet pressing following several steps in the processing, such as weighing, mixing in a ball mill, drying at room temperature, deagglomeration, screening, uniaxial pressing, isostatic pressing and sintering. SiC^1 dispersion in aqueous media has been conducted in a 1:1 ratio. To enhance mixing, this ratio is varied to 1:1,8, when using the SiC^2. For uniaxial pressing, a load of 40 MPA is used and for the isostatic pressing the load is increased to 300 MPa. Sintering is performed at 1900°C in an argon atmosphere.

The surface roughness of the ceramics is determined by means of a roughness MITUTOYO Surftest 301- point diamond. To analyze the average roughness (Ra) five samples per lot are used, with five measurements taken per sample. This parameter is used to compare the surface finish of the ceramic materials, produced with different types of SiC in the same processing conditions.

After the sintering process, the surface of the samples has been prepared for microscopic analysis and measurements of roughness, by the following sequence: grinding (SiC powder), machining (diamond tools) and polishing (diamond paste - 3μm).

After polishing, laser prints are made on the surfaces of samples SiC1 (CO_2 laser, SYNARD - FH Index Series, Marking Head), in order to improve the densification on the surface. To identify the best parameters for the use of the laser, we test various scan speeds (50 to 600 mm/s) and resolution (250-900 dpi), maintaining fixed power of 50 W. A second test is performed with a laser, making a coating prior before radiation, with slurry of SiC. In the mixture, ethanol is used as a solvent, we add a small amount of CMC, 0.07 g, (carboxymethyl cellulose), in order to agglomerate the particles and provide greater green resistance.

Diametral compression tests are carried out to evaluate the mechanical properties. The equipment used for testing is the Autograph AG-X SHIMATZU with load cell of 50 kN and speed of testing equal to 1 mm/min. An Epiphot 200 microscope (Nikon) coupled to a Zeiss camera - AxionCam ICC 3 is used to check the surfaces of ceramic samples. To observe details of the surface, a confocal laser microscope is also used.

RESULTS AND DISCUSSION

Figure 1 illustrates the changes in samples during processing steps: after pressing, after sintering and after machining and polishing.

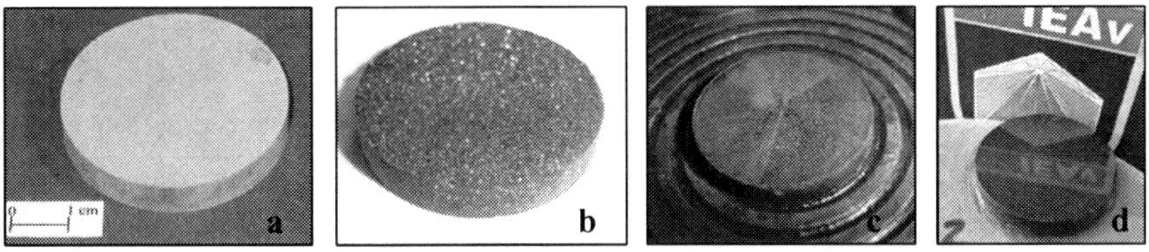

Figure 1 - Samples of SiC: a) before sintering, b) after sintering, c) grinded, d) polished sample.

Average roughness and apparent porosity

Table 1 shows the values of average roughness for SiC ceramics produced by pressing. Both groups, C^1 and C^2, show similar roughness values. The high value of roughness after sintering is associated with surface oxidation of the material and the presence of pores. The porosity seems to limit the quality of the surface finishing, becoming difficult to reduce to values lower than 1μm.

Table 1 - Average Roughness (Ra) of SiC ceramics samples.

Process	**Material**	**Average roughness of the samples (R_a)**	
		Only Sintered	**Machined and Polished**
C^1	SiC^1 e YAG (7,6%)	4,6± 1,3	1,33 ± 0,35
C^2	SiC^2 e YAG (7,6%)	4,7±0,8	1,41 ± 0,27

The apparent porosity, measured using the Archimedes method, shows a substantial difference between the samples prepared with the two types of silicon carbide. For SiC^1, we observe a higher porosity, 18.7 (± 1, 9), with a high fraction of interconnected pores, going from the top surface to bottom one. The other kind of carbide, SiC^2, produces no interconnecting pores. The apparent porosity shows a lower average value, 3.5 (± 0.8).

Diametral compression test

Ceramics made with SiC^1 (Raw Materials) show good mechanical strength for this type of test. The ceramics produced with SiC^2 (Starck), have a compressive strength of about 17.5% higher due to better densification in the samples. The Table 2 shows the results of these tests. The fracture characteristics are similar for both cases: all samples brake in traction, with small areas of fragmentation material; lack of crushing areas in the region of contact, absence of secondary cracks and lack of shear and compression fracture. Figure 2 shows a typical example of fracture for each condition studied.

Table 2 - Mechanical resistance of SiC ceramics subjected to diametral compression test.

Molding technique	Process	Material	Diametrical compressive strength (σ) – MPa
Isostatic pressing	C1	SiC[1] e YAG (7,6%)	120,0 ± 3,9
by the wet method	C2	SiC[2] e YAG (7,6%)	141,1 ± 2,8

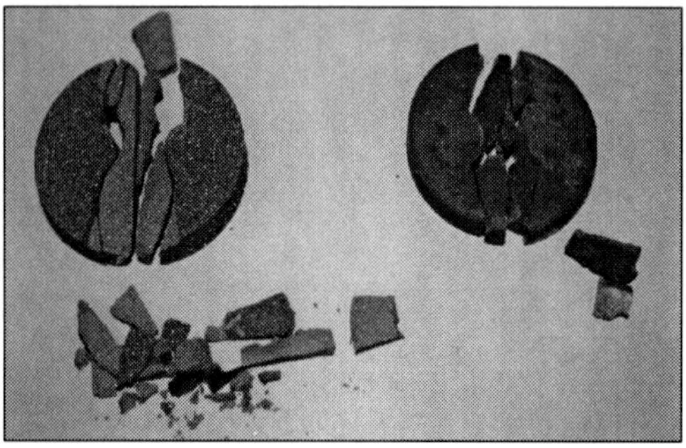

Figure 2 - Types of fractures in the samples: a) Raw Materials SiC, b) Starck SiC after diametral compression test.

Optical microscopy

The microscopic observations of polished surfaces, in samples with both types of silicon carbide, are quite similar, showing different pore sizes and distributed on the sample surface. Figure 3 illustrates this fact.

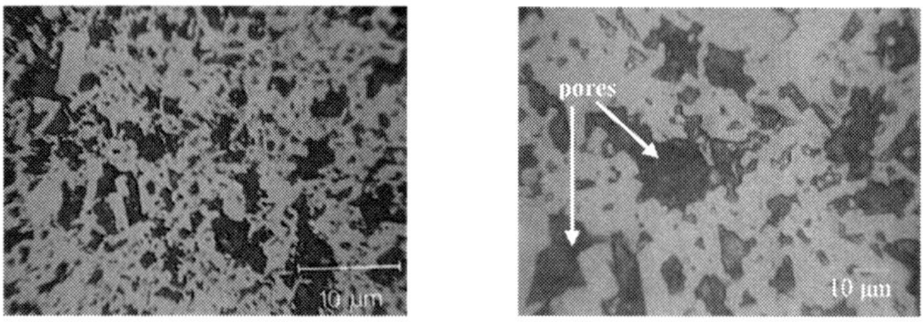

Figure 3 - Optical Microscopy (OM), showing pores on the surface of SiC samples

After irradiation of the surface using a laser directly on the sample, it is observed that for the most efficient parameters, partial melting occurrs and possible oxidation of the material, as shown in Figure 4. This phenomenon is observed for laser scan speeds of 50 and 100 mm/s at 600 and 900 dpi. Note in the images that the filling of the pores is not achieved as desired, rather an apparent sublimation of silicon takes place in the samples.

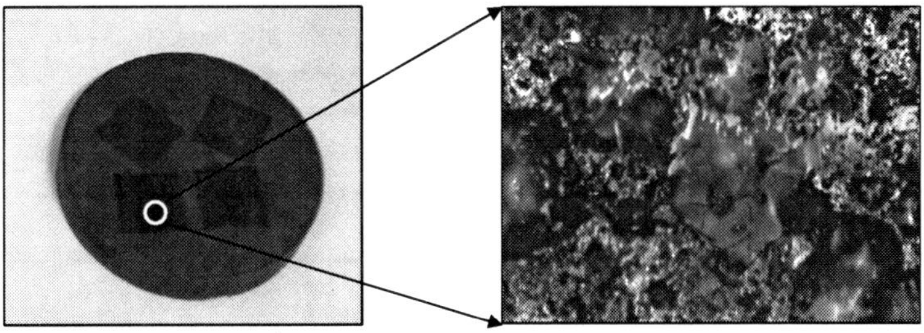

Figure 4 - Sample surface SiC^1 polished ceramic, after the application of laser without the use of a slurry of SiC: a) sample with radiation marks, b) micrograph of the irradiated region indicated.

Thereafter a new methodology is adopted i.e., the application of laser on a colloidal suspension and using the best possible laser parameters (speed - 50 mm/s; resolution - 600 dpi). Initial results with just one application of a layer of slip show a promising result, good adhesion and partial filling of pores. Figure 5 shows polished surfaces after irradiation with laser on the slurry of SiC.

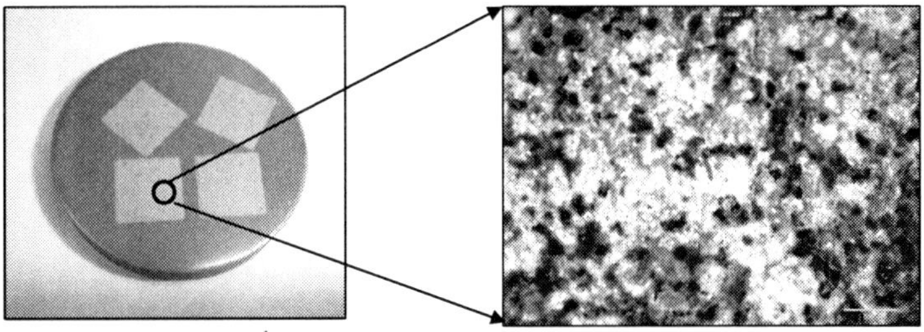

Figure 5 - Sample surface SiC^1 polished ceramic, after the application of laser with the use of a slurry of SiC: a) sample with radiation marks, b) micrograph of the irradiated region indicated.

CONCLUSIONS

1. The ceramics produced by the processes C^1 and C^2 have similar surface finishes, despite the higher porosity in the samples conformed to SiC Raw Materials (C^1);
2. The realization of the diametrical compression test shows that the samples produced by the processes C^1 and C^2 have similar modes of fracture: in traction, no traces of crushing at the contact points, absence of secondary cracks and areas of shear;
3. The use of laser on the sample surface without the use of slip seems to have caused oxidation and partial melting of the material, this occurrs for lower scan speeds (50 and 100 mm / s) and higher resolution (600 and 900 dpi). The use of laser on the sample surface after the application layer consists of ethanol, SiC and CMC allowed the sintering of a thin layer of SiC and contributes to reduction in size and amount of pores.

ACKNOWLEDGMENTS

We thank CNPq Projects 180978/2010-1, 183359/2009-7 and 183353/2009-9 and FINEP Project 555705/2009-9 for financial support.

REFERENCES

1. Dahotre, B. N., Kadolka P., Shah S., Surface and Interface Analysis, Vol. 31, 2001, pp. 659-672.
2. Izhevskyi, V. A., Genova, L. A., Bressiani, J. C., Bressiani, A. H. A., Materials Research Vol. 3, 2000, pp. 131-138.
3. Marins, E. M., 2008. "Otimização e caracterização microestrutural de cerâmica de carbeto de silício obtidas com material nacional para uso em blindagem balística", PhD. Thesis, Faculty of Engineering Campus Guaratinguetá, Universidade Estadual Paulista, Guaratinguetá, SP.
4. Pampuch, R., Journal of the European Ceramic Society, Vol. 18, 1998, pp. 994-995.

Mater. Res. Soc. Symp. Proc. Vol. 1373 © 2012 Materials Research Society
DOI: 10.1557/opl.2012.293

Synthesis and Characterization of Glass and Glass-Ceramic Materials of the System SiO_2-Fe_2O_3-BaO-Al_2O_3

O.I. Pánuco-Valdés, J. López-Cuevas, J.L. Rodríguez-Galicia
Centro de Investigación y de Estudios Avanzados del IPN, Unidad Saltillo, Carretera Saltillo-Monterrey, km 13.5, Ramos Arizpe, Coahuila, México, C.P. 25900

ABSTRACT

The effect of mechanical activation on the crystallization of glasses of the system SiO_2-Fe_2O_3-BaO-Al_2O_3 is investigated. The ceramic-glasses are synthesized from silica and iron-rich Ferro-Titanium-Zirconiferous (FTZ) sands available in the state of Coahuila, Mexico. The parent glass is subjected to mechanical activation in a high-energy attrition mill during 4 or 8h. Then, both milled and non-milled glasses are subjected to a thermal crystallization treatment at temperatures of 900 or 1000°C during 2 or 5h. It is found that, in the case of the thermal treatment carried out at 900°C/5h, an increment in the time employed for both the mechanical activation and the crystallization treatment enhanced the crystallization of the parent glasses and produced an increase in the compressive strength of the synthesized glass-ceramics.

INTRODUCTION

Iron is one of the elements present in large quantity in natural volcanic rocks; some basalts contain as much as 10 -15 wt% iron oxides; Ferro-Titanium-Zirconiferous (FTZ) sand contains ~15 wt% Fe_2O_3. Iron-rich glass-forming compositions are well known and widely used for the production of glass fibers and glass-ceramic tiles by melting, fibering or casting/rolling natural basalt rocks. Petrurgic products are characterized by high abrasion and corrosion resistance as well as high mechanical strength. Some of the iron-bearing silicate glasses also possess special optical and magnetic properties. In these glasses, iron can exist either in the ferrous or the ferric oxidation state. In the amorphous structure, Fe^{3+} can be considered as an intermediate ion, whereas Fe^{2+} is typically a modifying ion [1].

Controlled crystallization of glass is commonly employed to produce glass-ceramics with desirable properties. Glass-ceramic materials have a number of outstanding characteristics in comparison with traditional ceramics and glasses, which determine their application in advanced technology. The controlled crystallization of iron-rich glass is commonly employed to produce glass–ceramics with desirable properties [2]. Among the glass compositions that precipitate a magnetic phase are B_2O_3-Al_2O_3-SiO_2-Fe_2O_3-CaO, Li_2O-SiO_2-Fe_2O_3 and Li_2O-SiO_2-Fe_2O_3-MnO, BaO-B_2O_3-Fe_2O_3, SiO_2-Al_2O_3-Fe_2O_3-K_2O, CaO-Fe_2O_3-Al_2O_3-B_2O_3 and BaO-Fe_2O_3-SiO_2 [3]. The compositions with high iron content are peculiar since they show a tendency toward spontaneous crystallization and in some cases the amorphous structure can be obtained only after quenching in water [4].

The aim of this work is to synthesize and characterize glass and glass-ceramic materials with compositions in the SiO_2-Fe_2O_3-BaO-Al_2O_3 system and potential for structural applications. Silica and iron-rich FTZ sands available in the state of Coahuila, México are used as main raw material.

EXPERIMENTAL DETAILS

The raw materials used for the synthesis of the studied glass and glass-ceramic materials are: FTZ sand from General Cepeda, Coahuila, México, provided by the Mexican Geological Service; 99.43 wt% pure $BaCO_3$ (with a particle size of ~20μm, Alkem, México), and 97.19wt.% (with a particle size of ~51μm, Magnelec S.A. de C.V., México). The $BaCO_3$ and MgO are used as received, while the FTZ sand is purified (elimination of CaO) by lixiviation with acetic acid. The chemical composition of the conditioned FTZ sand is (wt.%): 50.29 SiO_2, 15.11 Fe_2O_3, 7.46 Al_2O_3, 7.40 Fe_2MgO_4, 6.47 TiO_2, 3.11 $ZrSiO_4$, 2.26%$BaSO_4$, 1 Na_2O, 0.58 K_2O, and 0.49 CaO. The mean particle size of this material is ~56μm.

The studied glass composition is (wt.%): 62.72 SiO_2, 13.15 Fe_2O_3, 7.0 TiO_2, 5.42 BaO, 4.88 MgO, 3.61 Al_2O_3, 1.48 ZrO_2, 0.84 Na_2O, 0.45 CaO, 0.23 K_2O, and 0.22 MnO. For the synthesis of this glass, the raw materials are mixed in the following proportions (wt.%): 81.66 FTZ sand, 16.70 $BaCO_3$ and 1.62 MgO. Then, the mixture is melted at 1450°C/2h in a Pt crucible using a Lindberg/blue M high temperature electric furnace. A decarbonation step is applied at 900°C/1h. The molten glass is quenched in ice water in order to preserve its amorphous state. Then the glass frit is milled for 5 min using a planetary ball mill with agate balls and containers, until a size of 150μm is obtained. This is followed by mechanical activation using a high-energy attrition mill for 4 or 8h, employing MgO-partially-stabilized ZrO_2 balls as milling media, with a mass ratio of 5:1 with respect to the material loaded into de mill, and using ethylic alcohol as dispersion medium. The initial particle size of the glass powder is ~55μm, which is reduced to ~47μm and ~27μm after 4 and 8h of mechanical activation, respectively. Then, the glass powder is uniaxially pressed at 4 Ton in order to obtain cylinders with a diameter of 1.2cm and a height of 0.6cm. These cylinders are heat treated at temperatures of 900 or 1000°C during 2 or 5h in order to obtain the desired glass-ceramic materials, using a Thermolyne 46200 high temperature electric furnace. A heating rate of 5°C/min and cooling rate of 10°C/min are used.

Characterization by X-Ray Diffraction (XRD) is performed to evaluate the crystallization process and phase composition of the synthesized glass-ceramics. XRD is performed in a Philips X'Pert 3040 apparatus, with CuKα radiation (λ = 1.54 Å) at an accelerating voltage of 40 kV and current of 30 mA. SRD scans are carried out from 10 to 80° 2θ, and a step size of 0.03° (2θ/s). The glass-ceramic samples are analyzed also by Scanning Electron Microscopy (SEM) using a Phillips XL30 ESEM operated at an accelerating voltage of 20 KV, backscattered electron imaging (BEI) mode and working distance of 10 mm.

Finally, the compressive strength of the synthesized glass-ceramics is determined by using a CONTROLS press fitted with a 15 KN cell, with a load application rate of 10 MPa/s. Three tests are carried out per each synthesized glass-ceramic material; the average values are reported in this work.

DISCUSSION

Phase composition of the synthesized glass-ceramics

The XRD patterns obtained for the synthesized glass-ceramics are shown in Figures 1 and 2, which correspond to crystallization temperatures of 900 and 1000°C, respectively.

According to Figure 1, at 900°C/2h the crystalline phases formed for the three milling times employed are: Hematite (Fe_2O_3), Cristobalite (SiO_2) and Celsian ($BaAl_2Si_2O_8$, with monoclinic crystalline structure). The intensity of the reflections of these three phases increased with increasing milling time, which is more noticeable for the case of the first two phases. Thus, in a qualitative sense, it can be argued that mechanical activation of the parent glass makes it more prone to crystallize at this treatment temperature. In all cases, Hematite and Cristobalite are the main crystalline phases formed, while Celsian had the lowest proportion in the glass-ceramic materials. The fact that the formation of Celsian took place even for the case of the material without mechanical activation, suggests that under the conditions studied this phase is more stable than the hexagonal polymorphic $BaAl_2Si_2O_8$ phase, known as Hexacelsian, which is not detected. This is an advantage, since the latter phase is undesirable due to its relatively high thermal expansion coefficient and because it undergoes a conversion into an orthorhombic phase at ~300°C, which is associated with a volume change of ~3-4%. This causes microcracking in the material, negatively affecting its final mechanical properties.

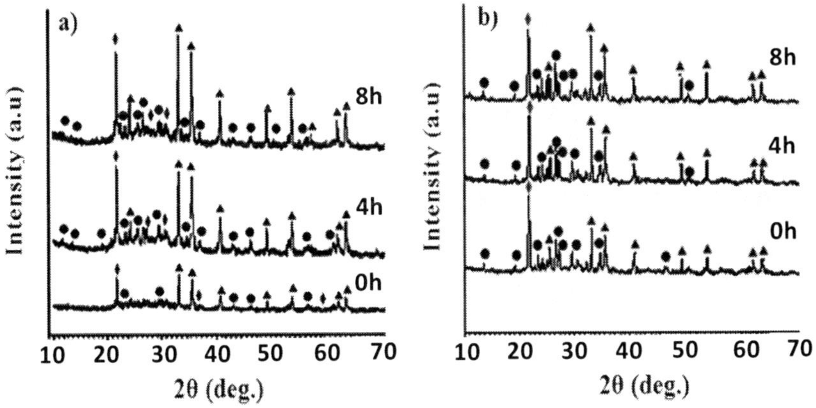

Figure 1. XRD patterns obtained for the glass-ceramic materials synthesized at 900°C, with holding times of: a) 2h and b) 5h. Parent glasses mechanically activated for 0-8h. ▲ Hematite (Fe_2O_3), ◆ Cristobalite (SiO_2) and ● Celsian ($BaAl_2Si_2O_8$).

The same crystalline phases are observed at 900°C/5h. However, in this case with increasing milling time the intensity of the reflections of Cristobalite remained nearly unchanged, while the intensity of the reflections of Hematite and Celsian increased slightly. For all glass-ceramic materials synthesized under these conditions, Cristobalite and Hematite are the main crystalline phases formed while Celsian is present in the lowest proportion. Nevertheless, this is higher than that observed for the shorter holding time employed.

The same crystalline phases observed at 900°C are found in the glass-ceramics synthesized at 1000°C, Figure 2, but with the additional formation of pseudobrookite (Fe_2TiO_5) in the latter case. This is attributed to an enhanced chemical reactivity of the mechanically activated glass at the higher crystallization temperature [5]. However, independently of the holding time employed at 1000°C, at this temperature the intensity of the reflections of all crystalline phases detected is relatively unaffected by the milling time used. In all cases, Hematite, Cristobalite and pseudobrookite are the main crystalline phases formed, while Celsian had the lowest proportion in the glass-ceramic materials.

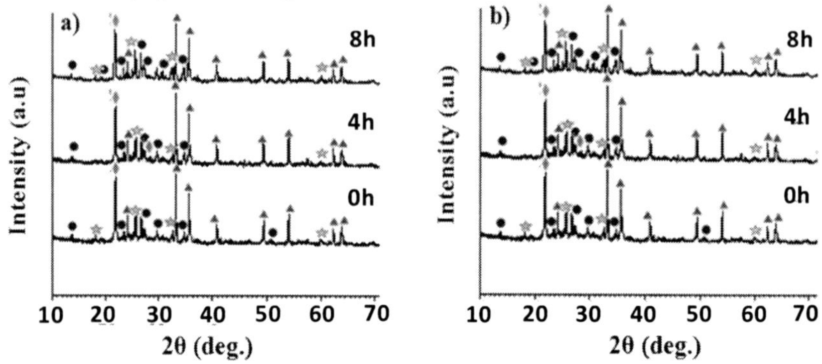

Figure 2. XRD patterns obtained for the glass-ceramic materials synthesized at 1000°C, with holding times of: a) 2h and b) 5h. Parent glasses mechanically activated for 0-8h. ▲ Hematite (Fe_2O_3), ◆ Cristobalite (SiO_2), ● Celsian ($BaAl_2Si_2O_8$), and ★ Pseudobrookite (Fe_2TiO_5).

Microstructure of the synthesized glass-ceramics

Figure 3 shows the SEM micrographs obtained for the synthesized glass-ceramics. It can be observed that the formed crystals became finer and more homogeneously distributed in the microstructure of the materials when the time of mechanical activation of the parent glasses and the crystallization temperature and time are increased.

Compressive strength of the synthesized glass-ceramics

As can be seen in Table I, the compressive strength of the glass-ceramics synthesized at 900°C increased with increasing time employed for both the mechanical activation and the crystallization treatments. This can be attributed to the higher crystallization degree, as well as to the higher proportion of some crystalline phases such as Celsian, attained in the glass-ceramic materials synthesized under such conditions. Two additional factors further contributing to this are the finer size and more homogeneous distribution achieved by the crystals in the microstructure under the mentioned conditions. In this case, the maximum compressive strength obtained is 155.7 ±5 MPa, which corresponded to the glass-ceramic synthesized at 900°C/5h with 8h of milling of the parent glass.

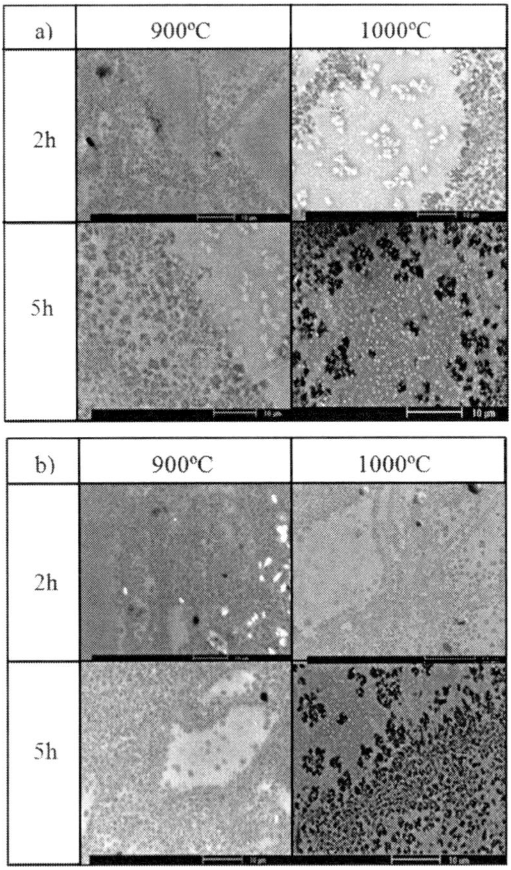

Figure 3. SEM BEI micrographs obtained for the glass-ceramics synthesized using parent glasses mechanically activated for: a) 0h and b) 8h, as a function of time and temperature of the crystallization treatment.

When the synthesis temperature is increased from 900 to 1000°C, for milling times of the parent glass of up to 4h, the compressive strength of the synthesized glass-ceramics tended to increase with an increment in the crystallization temperature at a given holding time. However, the opposite trend is observed with 8h of milling. On the other hand, at a constant crystallization temperature of 1000°C, again for milling times of up to 4h, the compressive strength of the glass-ceramics decreased with increasing time of crystallization. In contrast, increasing the milling time to 8h resulted in the opposite trend. However, the compressive strength values obtained by using the latter milling time are much smaller than those obtained with the other two milling times. In this case, the maximum compressive strength obtained is 137.4 ± 2.65 MPa, which corresponded to the glass-ceramic synthesized at 1000°C/2h with 4h of milling. As can be seen, this value is smaller than the maximum value obtained for the glass-ceramic synthesized at 900°C/5h with 8h of milling. These results suggest that the presence of pseudobrookite in the samples crystallized at 1000°C appears to be detrimental for their mechanical properties.

Table I. Compressive strength of the synthesized glass-ceramics (MPa).

Crystallization temperature (°C)	Milling and crystallization time (h)					
	0		4		8	
	2	5	2	5	2	5
900	72 ±6.5	125.6 ±3.8	72.9 ±10.5	134.2 ±4	121.2 ±.17	154.7 ±3.8
1000	136.8 ±2	125.0 ±10.11	137.4 ±2.65	134.9 ±7.36	75.2 ±3.2	82.2 ±2

CONCLUSIONS

For crystallization of the parent glass at 900°C/2h, the amount of the formed phases [Hematite (Fe_2O_3), Cristobalite (SiO_2) and Celsian ($BaAl_2Si_2O_8$)] increased with increasing milling time. However, at 900°C/5h the relative proportions of the phases are only slightly affected by this factor. At 1000°C, pseudobrookite (Fe_2TiO_5) is additionally formed, with the relative proportions of the crystalline phases being relatively unaffected by both milling and holding times employed. At 900°C, the compressive strength increased with increasing milling and holding times, with a maximum value of 154.7 ± 3.8 MPa obtained at 900°C/5h with 8h of milling. The presence of pseudobrookite in the samples crystallized at 1000°C seems to be detrimental for their mechanical properties.

ACKNOWLEDGMENTS

The authors express their gratitude to CONACYT and Cinvestav-Saltillo for the financial support and facilities provided for the development of this work.

REFERENCES

1. A. Karamanov, P. Pisciella, C. Cantalini and M. Pelino, *J. Am. Ceram. Soc.* **83**, 3153-3157 (2000).
2. A.A. Francis, *Mater. Res. Bull.* **41**, 1146-1154 (2006).
3. A.A. Francis, *Ceram. Int.* **33**, 163-168 (2007).
4. A. karamanov and M. Pelino, *J. Non-Cryst. Solids* **281**, 139-151 (2001).
5. A.A. Cristóbal, E.F.Aglietti and J.M. Porto López, *SAM/CONAMET **2007***, San Nicolás, Argentina, 1021-1026 (September 4-7, 2007).

Mater. Res. Soc. Symp. Proc. Vol. 1373 © 2012 Materials Research Society
DOI: 10.1557/opl.2012.294

Effect of Milling Time on the Physical and Mechanical Properties of Celsian-Mullite Composites Synthesized from Coal Fly Ash

Jorge López-Cuevas[1], David Long-González[2], Carlos A. Gutiérrez-Chavarría[1]
[1]Cinvestav-Saltillo, Carr. Saltillo-Monterrey, km 13.5, Ramos Arizpe, Coahuila, México, 25900
[2]RHI REFMEX, S.A. de C.V., Carr. Saltillo-Monterrey, Km. 9, Ramos Arizpe, Coahuila, México, 25900

ABSTRACT

Four Celsian ($Ba_{0.75}Sr_{0.25}Al_2Si_2O_8$)/Mullite ($Al_6Si_2O_{13}$) composites, with potential structural applications at high temperatures, are synthesized from coal fly ash (byproduct of a Mexican coal-burning power plant, constituted mainly by SiO_2 and Al_2O_3). Nominal Celsian/Mullite weight ratios studied are 80/20, 60/40, 40/60 and 20/80. Mullite is synthesized separately at 1600°C/2h and then mixed with a Celsian precursor mixture previously calcined at 900°C/5h. During this process the Celsian phase is formed by a solid state reaction at 1100-1400°C/5h. Prior to this, the reacting mixture is milled in a planetary mill during 1 or 2h and then compacted by uniaxial and cold isostatic pressing. The microstructure and phase composition of the synthesized composites are characterized by X-Ray Diffraction (XRD) and Scanning Electron Microscopy (SEM/EDS). Their dynamic Young's modulus is measured by an ultrasonic technique, and their mechanical strength is evaluated from flexural tests carried out at room temperature. The expected phases are obtained in all cases, although with some differences with respect to their expected relative proportions, according to the studied nominal compositions. In general, the longest milling time employed produced samples with the largest degree of crystallinity and density, as well as with the best microstructural characteristics and mechanical properties.

INTRODUCTION

During the last ten years, ceramic composites based on barium aluminosilicate (BAS, $BaAl_2Si_2O_8$) have attracted a great deal of attention for structural applications at high temperatures. The BAS phase, especially in its monoclinic polymorphic form known as Celsian or Monocelsian, possesses good mechanical properties, low thermal expansion coefficient, high resistance to oxidation, high resistance to reduction and high resistance to attack by molten slags. It also has stable dielectric properties and good resistance to thermal shock [1-3]. Celsian is highly refractory due to its high melting point (1760°C), it is stable below 1590°C and it is useful as a matrix for ceramic composites [4,5]. BAS also has a hexagonal polymorphic form known as Hexacelsian, which tends to form before Celsian and frequently remains in a metastable state at low temperatures [6]. Hexacelsian is undesirable due to its relatively high thermal expansion coefficient and because it undergoes a conversion into an orthorhombic form at ~300°C, which is associated with a volume change of ~3-4%. This causes microcracking in the material, negatively affecting its final mechanical properties [3,7,8]. Partial substitution of BaO by SrO in BAS promotes the formation of $Ba_{1-x}Sr_xAl_2Si_2O_8$ solid solutions with monoclinic crystalline structure [9], which are denominated either as barium-strontium Celsian or simply as Celsian. In contrast, the hexagonal $Ba_{1-x}Sr_xAl_2Si_2O_8$ solid solutions are denominated as Hexacelsian. Thus, this can be considered to be an effective way to promote the Hexacelsian to Celsian conversion. The stabilization of Celsian can be achieved also by small additions of mineralizing agents such as Li_2O, MgO, CaO, Na_2O and TiO_2 [10], as well as by manipulation of the processing conditions [11]. A topic that has not been studied in detail is the stabilization of Celsian when this phase is synthesized *in situ* as the matrix of ceramic composites.

In this work, we investigate the *in situ* synthesis of Celsian ($Ba_{0.75}Sr_{0.25}Al_2Si_2O_8$) - Mullite ($Al_6Si_2O_{13}$) composites by solid state reaction, using Coal Fly Ash (byproduct of a Mexican coal-burning power plant, constituted mainly by SiO_2 and Al_2O_3) as main raw material. The main objective of this work is to study the dependence of the physical and mechanical properties of the synthesized composite materials with respect to their composition and milling time of the initial raw materials.

EXPERIMENTAL DETAILS

Composite compositions

The studied nominal Celsian/Mullite composite compositions (in wt.%) and denominations used (given in parentheses) are: (1) 80/20 (8C2M), (2) 60/40 (6C4M), (3) 40/60 (4C6M), and (4) 20/80 (2C8M). According to the Celsian-Mullite compatibility join reported by Semler and Foster [12], these two phases are compatible, and thus, are able to coexist in equilibrium below 1554°C.

Raw materials

The raw materials employed in the present investigation are as follows:

1) Coal Fly Ash (CFA) from the "José López Portillo" power plant (Nava, Coahuila, México). This material is milled in a ball mill during 5h and subject to a wet magnetic separation process to eliminate as much iron oxide as possible. The chemical composition of the final product is (in wt.%): 64.59% SiO_2, 25.73% Al_2O_3, 2.57% Fe_2O_3, 2.12% CaO, 0.59% TiO_2, 0.48% MgO, 0.46% Na_2O, 0.37% K_2O, and 1.11% of other oxides (SrO, ZrO_2, CuO, NiO, SO_3 and P_2O_5). Ignition losses (free C) are 1.94% and the final powder has a mean particle size of ~5 μm, containing quartz, mullite and glass as main phases
2) $BaCO_3$, with a purity of 99.43 wt.% and mean particle size of ~3 μm (Alkem, México)
3) $SrCO_3$, with a purity of 97.86 wt.% and mean particle size of ~4 μm (Solvay, México)
4) Al_2O_3, with a purity of 99.99 wt.% and mean particle size < 1 μm (HPA-0.5 Sasol, USA).

Procedure

Stoichiometric Mullite is synthesized at 1600°C/3h from a suitable mixture of the raw materials. Then, the synthesized Mullite is milled in a two-station Retsch PM-200 planetary mill during 1h at 350rpm using agate balls and mortars. A powder to balls weight ratio of 1:2 is used. Following this, a suitable mixture of raw materials is prepared for the synthesis of Celsian with the desired composition. This mixture is calcined at 900°C/5h, giving rise to the formation of a mixture of barium silicates and barium aluminate. The calcined material is then mixed with previously synthesized and milled Mullite in the pre-established proportions. This mixture is then milled during 1 or 2h employing the same planetary mill and milling conditions used before.

The milled mixtures are uniaxially pressed at 30 MPa, followed by cold isostatic pressing at 200 MPa in order to obtain cylinders with a diameter of ~1.2 cm and a height of ~ 0.5 cm. The cylinders are subjected to solid state reaction during 5h at 1100-1400°C in order to promote the *in situ* synthesis of the desired ceramic composites. The agreement between the analyzed and the nominal chemical composition of all synthesized materials is verified. The main difference found between nominal and actual chemical compositions is due to the presence of small amounts of Fe_2O_3 and CaO impurities coming from the CFA. The apparent density ($\rho_{apparent}$) of the sintered cylinders is determined by using the Archimedes' principle in water, and their theoretical density ($\rho_{theoretical}$) is calculated from the densities of $Ba_{0.75}Sr_{0.25}Al_2Si_2O_8$ and Mullite taken from the corresponding JCPDS cards, as well as by using the rule of mixtures. The relative density ($\rho_{relative}$) of the synthesized composites is determined using equation 1:

$$\rho_{relative} = (\rho_{apparent} / \rho_{theoretical}) \times 100 \qquad (1)$$

Phase analysis of the sintered samples is carried out by XRD using a Siemens D-5000 diffractometer employing CuKα radiation in the 2θ range of 10-70°; a step size of 0.05° and a counting interval of 3s are used. Microstructural analysis of the sintered samples is carried out by SEM using a Zeiss DSM 950 instrument coupled to an EDS system for the microanalysis of phases.

The dynamic Young's modulus of the sintered samples is evaluated on 3 mm-thick plates by the pulse-echo ultrasonic technique employing a Panametrics-NDT™ EPOCH 4 apparatus and using equation 2:

$$E = \frac{V_L^2 \rho}{1.499} \qquad (2)$$

where E = Young's modulus (GPa), V_L = longitudinal speed of the ultrasonic wave through the material (m/s), and ρ = apparent density of the material (Kg/m^3).

Three-point bending strength is measured at room temperature using a MICROTEST universal testing machine. In this case, sintered bars with dimensions 3x4x50 mm and a supporting span length of 30 mm are tested at a loading rate (crosshead speed) of 0.5 mm/min.

DISCUSSION

Phase composition of synthesized composites

Table 1 gives a summary of the effect of milling time and treatment temperature on the phase compositions determined by XRD for the synthesized Celsian/Mullite composites. Two examples of the obtained XRD patterns are shown in Figures 1 and 2. These XRD patterns correspond to the 8C2M and 2C8M composites, respectively, after milling during 1 or 2h, followed by heat treatment at 1100-1400°C/5h.

Table 1. Summary of phase composition of synthesized Celsian/Mullite composites. Key: C = monoclinic $Ba_{0.75}Sr_{0.25}Al_2Si_2O_8$ (Celsian), H = hexagonal $Ba_{0.75}Sr_{0.25}Al_2Si_2O_8$, (Hexacelsian), M = Mullite, BA = $BaAl_2O_4$, A = Al_2O_3, and t = traces.

8C2M composite		
Temperature (°C)	Milling time (h)	
	1	2
1100	C,M,A^t,BA^t	C,M,H,A
1200	C,M,A,BA^t	C,M,A
1300	C,A,M,BA^t	C,A,M
1400	C,A,M^t,BA^t	C,A,M^t
6C4M composite		
Temperature (°C)	Milling time (h)	
	1	2
1100	C,M,H,A^t	C,M,H
1200	C,M,A	C,M,A
1300	C,M,A	C,M,A
1400	C,M,A	C,A,M^t
4C6M composite		
Temperature (°C)	Milling time (h)	
	1	2
1100	M,C,H,A^t	M,C,H,A
1200	M,C,A^t	M,C,A
1300	C,M,A^t	C,M,A
1400	C,M,A	C,M,A
2C8M composite		
Temperature (°C)	Milling time (h)	
	1	2
1100	M,H,C,A	M,H,C,A
1200	M,C,A	M,C,A,H^t
1300	M,C,A	M,C,A
1400	M,C,A	M,C,A

In general, milling during 1h, for all studied compositions and treatment temperatures, results in formation of the desired Celsian and Mullite phases in the synthesized materials. However, the additional presence of some unexpected secondary phases (Hexacelsian, Al_2O_3 and $BaAl_2O_4$, plus a glassy phase) is also observed. Furthermore, the amount of Mullite present in the synthesized composites is considerably lower than the desired one. This is probably due to the reaction of the pre-existing Mullite with the Celsian precursors present in the mixture, giving rise to the formation of Al_2O_3, $BaAl_2O_4$ and glassy phase, according to the following thermodynamically feasible chemical reactions:

$$Al_6Si_2O_{13} + 2BaSiO_3 = 2BaAl_2Si_2O_8 + Al_2O_3 \qquad \Delta G^{1100°C} = -29.27\ \text{kcal}$$
$$3Al_6Si_2O_{13} + 2Ba_2SiO_4 = 4BaAl_2Si_2O_8 + 5Al_2O_3 \qquad \Delta G^{1100°C} = -41.72\ \text{kcal}$$
$$Al_6Si_2O_{13} + 3BaSi_2O_5 = 3BaAl_2Si_2O_8 + 2SiO_2 \qquad \Delta G^{1100°C} = -33.96\ \text{kcal}$$

$$Al_6Si_2O_{13} + BaAl_2O_4 = 3Al_2O_3 + BaAl_2Si_2O_8 \qquad \Delta G^{1100°C} = -67.88 \text{ kcal}$$

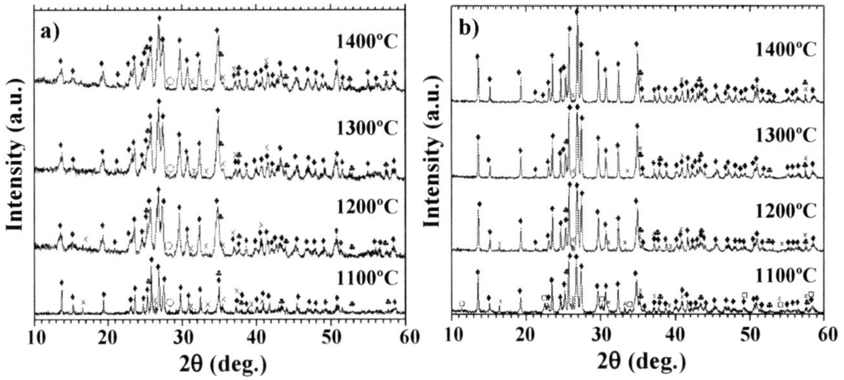

Figure 1. XRD patterns of 8C2M composites milled for: a) 1h and b) 2h, and heat treated at 1100-1400°C/5h. Key: ♦ Celsian, x Mullite, □ Hexacelsian, ♣ Al_2O_3 and ○ $BaAl_2O_4$.

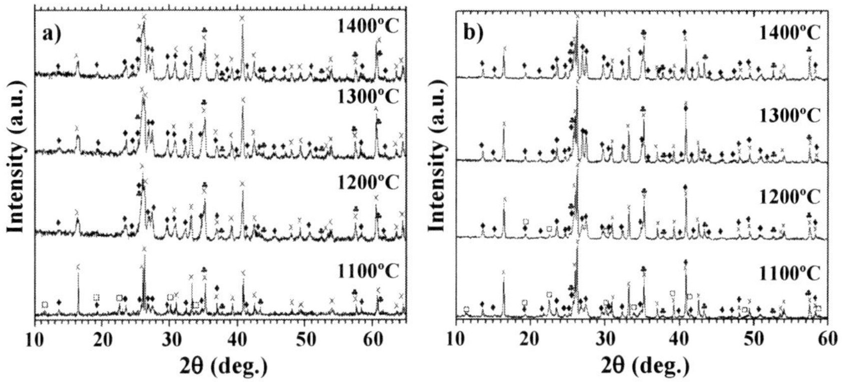

Figure 2. XRD patterns of 2C8M composites milled for: a) 1h and b) 2h, and heat treated at 1100-1400°C/5h. Key: ♦ Celsian, x Mullite, □ Hexacelsian and ♣ Al_2O_3.

The formation of the glassy phase detected in the synthesized composites is probably promoted by the presence of large amounts of alkaline-earth oxides (CaO, BaO and SrO) and of impurities such as Fe_2O_3 in the Celsian precursors. These oxide impurities have a fluxing effect that resulted in the generation of a great deal of liquid at high temperatures [13]. These oxides also increased the mullitization temperature [14]. This means that in the presence of oxide impurities Mullite is stable only at temperatures higher than those employed for the synthesis of the composites, causing the consumption of part of the pre-existing Mullite at the latter temperatures. Further liquid formation and Mullite consumption could arise from the ternary eutectic reaction that can occur at ~1300°C between SiO_2, Celsian and Mullite [15]. Besides, Mullite can also be dissolved by liquid formed during sintering of the material [16-18].

It is observed that the $BaAl_2O_4$ phase is formed only in the composite composition (8C2M) with the largest content of Al_2O_3, for which the degree of reaction is the least. However, Al_2O_3 appears in all compositions and treatment temperatures employed, and Hexacelsian is detected in the three compositions having the highest Mullite content (6C4M, 4C6M and 2C8M), when these are heat treated at the lowest treatment temperature employed (1100°C). Since it is known that Hexacelsian is surface nucleated [19], it is probable that, with increasing Mullite content in the composites, the number of sites for the surface nucleation of Hexacelsian increased, with a consequent increase in the amount of this phase. However, it is noticed that the amount of Hexacelsian decreases in the composite materials with increasing temperature, which is evidently due to its conversion into Celsian. Based on these results, it can be said that the behavior of the materials milled during 1h is dominated by local heterogeneities present in the initial mixtures, which results in the formation of non-equilibrium phases.

For the case of milling during 2h, for all studied compositions and treatment temperatures, the XRD reflections of all formed crystalline phases become more intense and well-defined, indicating that a higher degree of crystallization is attained in the synthesized materials. The detected phases are: Celsian, Mullite, Hexacelsian and Al_2O_3; $BaAl_2O_4$ and glassy phase are not found. This is probably due to the fact that the initial mixtures are more homogeneous due to the longer treatment time employed, which results in more complete chemical reactions. However, Hexacelsian appears in more cases when compared with the materials milled during 1h. Probably, when a longer milling time is used, it results in a larger number of sites for the surface nucleation of Hexacelsian. Again, the amount of this phase decreases with increasing temperature, due to its conversion into Celsian. With respect to the Celsian and Mullite phases detected in the synthesized composites, the XRD patterns show again that the amount of Mullite is considerably lower than the desired one. It also shown that the relative proportions of Celsian and Al_2O_3 increase, while that of Mullite decreases, with increasing treatment temperature. It is believed that this is due to the occurrence of some chemical reactions similar to those mentioned above.

Microstructure of synthesized composites

In general, the SEM analyses corroborate the results obtained from the XRD studies. Figure 3 shows the microstructure observed on the SEM for 8C2M and 2C8M composites synthesized using a milling time of 1h and a sintering temperature of 1400°C. In both cases, the samples seem to be biphasic, with the presence of a considerable amount of porosity, with a pore size of 6-30 μm for the case of the first material, and of 8-40 μm for the case of the second one. Porosity is ~11% and ~9%, respectively.

From the results of the EDS analysis, it is deduced that the clear and dark phases observed in the 8C2M composite (Figure 3a) are Celsian and alumina, respectively; Mullite is not found in this sample. Similarly, the clear and dark phases observed in the 2C8M composite (Figure 3b) correspond to Celsian and Mullite, respectively; alumina is not observed in this sample, probably due to its small proportion in the material. Although a glassy phase is not clearly distinguished on the SEM in neither case, its presence is evidenced by a severe attack occurring in some of the samples during HF-etching. Observation on the SEM reveals the absence of a substantial amount of cracks in the analyzed samples. When the composites are synthesized using a milling time of 2h, the corresponding microstructures are similar to those shown in Figure 3. However, it is observed that porosity slightly decreases with increasing milling time.

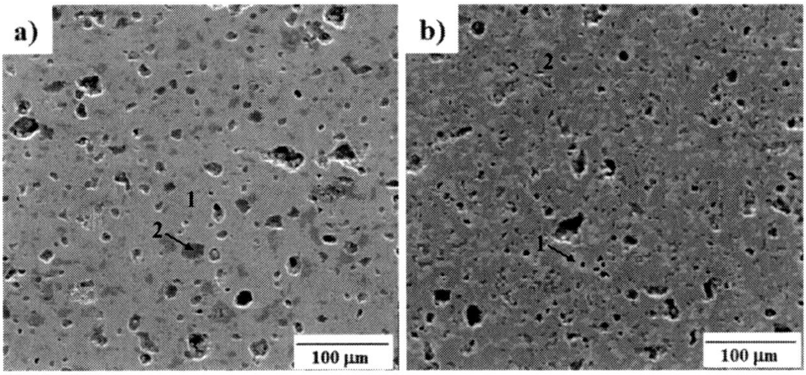

Figure 3. SEM micrographs of composites synthesized using a milling time of 1h and sintering at 1400°C: a) 8C2M composite (phase 1 = Celsian; phase 2 = Al_2O_3), and b) 2C8M composite (phase 1 = Celsian; phase 2 = Mullite).

Density of synthesized composites

As can be seen in Table 2, Mullite has a density lower than that of Celsian. Therefore, the theoretical density of the composites should decrease with increasing nominal Mullite content, independently of the milling time employed. However, according to Tables 3 to 6, the opposite behavior is observed for the case of the experimental density values. It is clear that both the apparent and the relative densities of the synthesized composites tend either to increase or to remain relatively constant with increasing nominal Mullite content, for all milling times and treatment temperatures employed. This is due to the fact that, for all nominal composite compositions and milling times under investigation, the actual Mullite content of the synthesized materials is considerably lower than the desired one, with a concomitant increment in the relative proportions of other phases such as Celsian and Al_2O_3, which have densities that are higher than that of Mullite. This is accentuated by increasing treatment temperatures. Tables 3 to 6 also show that the samples synthesized using a milling time of 2h achieved slightly higher apparent and relative densities with respect to those achieved by the materials that are milled during 1h. This effect is attributed to a larger amount of Celsian and Hexacelsian together with a smaller amount of porosity achieved in the first case.

Table 2. Theoretical densities of Celsian, Mullite, and studied nominal composites.

$\rho_{theoretical}$ (g/cm^3)					
Celsian (JCPDS card no. 038-1451)	8C2M	6C4M	4C6M	2C8M	Mullite (JCPDS card no. 079-1450)
3.314	3.27	3.23	3.18	3.14	3.103

Table 3. Apparent and relative densities of composites synthesized with 1h of milling and heat treatment at 1350°C (6C4M and 4C6M) or 1400°C (8C2M and 2C8M) for 5h.

Composite	$\rho_{apparent}$ (g/cm^3)	$\rho_{relative}$ (%)
8C2M	2.89±0.01	88.3±0.3
6C4M	2.86±0.01	88.6±0.3
4C6M	2.92±0.01	91.9±0.1
2C8M	2.98±0.01	94.8±0.1

Table 4. Apparent and relative densities of composites synthesized with 2h of milling and heat treatment at 1300°C/5h.

Composite	$\rho_{apparent}$ (g/cm^3)	$\rho_{relative}$ (%)
8C2M	2.95±0.02	90.1±0.8
6C4M	2.94	91.1
4C6M	2.99±0.01	93.9±0.3
2C8M	2.93±0.03	93.2±0.8

Table 5. Apparent and relative densities of composites synthesized with 2h of milling and heat treatment at 1350°C/5h.

Composite	$\rho_{apparent}$ (g/cm^3)	$\rho_{relative}$ (%)
8C2M	2.97±0.01	90.8±0.2
6C4M	3.03±0.01	94.0±0.5
4C6M	3.03±0.01	95.3±0.5
2C8M	3.0±0.01	95.6±0.5

Mechanical properties of synthesized composites

Tables 7 and 8 show that, for both milling times employed, the mechanical properties of the synthesized composites increase with increasing nominal Mullite content. As already discussed, for all nominal composite compositions, milling times and treatment temperatures employed, the actual Mullite content of the synthesized materials is considerably lower than the desired one, with a concomitant increment in the relative proportions of other phases such as Celsian and Al_2O_3. Thus, it is the presence of the latter phases, as well as the higher degree of densification achieved by the synthesized materials with increasing sintering temperature, the factors truly responsible for the mechanical behavior of the synthesized composites. However, the samples synthesized using a milling time of 2h achieved higher mechanical properties with respect to the materials synthesized using a milling time of 1h, which is probably due to a smaller amount of porosity in the first case.

Table 6. Apparent and relative densities of composites synthesized with 2h of milling and heat treatment at 1400°C/5h.

Composite	$\rho_{apparent}$ (g/cm^3)	$\rho_{relative}$ (%)
8C2M	3.0	91.76
6C4M	3.04±0.01	94.3±0.2
4C6M	3.04±0.01	95.4±0.2
2C8M	3.0±0.01	95.6±0.2

Table 7. Mechanical properties of composites synthesized using a milling time of 1h and sintering at 1350°C (6C4M and 4C6M) or 1400°C (8C2M and 2C8M) for 10h.

Composite	Three-point flexural strength (MPa)	E (GPa) from ultrasonic testing	E (GPa) from mechanical testing
8C2M	95±3	56±0.5	49.8±4.0
6C4M	109±4	56±2.0	54.9±1.5
4C6M	128±6	72±0.5	63±4.0
2C8M	123±8	75±1.0	64.6±2.6

Table 8. Mechanical properties of composites synthesized using a milling time of 2h and sintering at 1350°C (6C4M and 4C6M) or 1400°C (8C2M and 2C8M) for 10h.

Composite	Three-point flexural strength (MPa)	E (GPa) from ultrasonic testing	E (GPa) from mechanical testing
8C2M	120.6±7.0	73.3±0.1	62.0±0.5
6C4M	146.4±12.7	71.7±0.1	65.1±5.9
4C6M	111.1±8.5	73.0±1.5	61.2±4.2
2C8M	166.2±14.4	100.3±1.8	77.5±2.8

CONCLUSIONS

In all cases, the Mullite content of the synthesized composites is considerably lower than the nominal one. This is due to the reaction of Mullite with the Celsian precursors present in the initial mixture, which results in the formation of Al_2O_3, $BaAl_2O_4$ and glassy phases. Other important factors contributing to the consumption of Mullite are: (1) Presence of CaO, BaO, SrO and Fe_2O_3 in the Celsian precursors, all of which had a fluxing effect and increased the mullitization temperature, and (2) Occurrence of a ternary eutectic reaction at ~1300°C between SiO_2, Celsian and Mullite. The surface nucleation of Hexacelsian is promoted by increasing both the Mullite content and the milling time. However, the amount of Hexacelsian decreased with increasing temperature due to its conversion into Celsian. The apparent and relative densities, as well as the mechanical properties of the composites, increase with increasing Mullite content. From the results obtained, it is apparent that a key factor responsible for this effect is the higher degree of densification achieved by these materials under the processing conditions investigated.

ACKNOWLEDGMENTS

The authors express their gratitude to CONACYT and Cinvestav-Saltillo for the financial support and experimental facilities provided for the development of this work.

REFERENCES

1. X.-D. Zhang, K.H. Sandhage and H.L. Fraser, *J. Am. Ceram. Soc.* **81**, 2983-2997 (1998).
2. R.A. McCauley, *J. Mater. Sci.* **35**, 3939-3942 (2000).
3. J.S. Moya Corral and G. Verduch, *Brit. Ceram. Trans. J.* **77**, 40-44 (1978).
4. V.S.R. Murthy and M.H. Lewis, *Brit. Ceram. Trans. J.* **89**, 173-174 (1990).
5. N.P. Bansal, *J. Am. Ceram. Soc.* **80**, 2407-2409 (1997).
6. H.C. Lin and W.R. Foster, *Am. Mineral.* **53**, 134-144 (1968).
7. B. Yoshiki and K. Matsumoto, *J. Am. Ceram. Soc.* **80**, 2021-2029 (1997).
8. N. Frety, A. Taylor and M.H. Lewis, *J. Non-Cryst. Solids* **195**, 28-37 (1996).
9. N.P. Bansal, M.J. Hyatt and C.H. Drummond III, *Ceram. Eng. Sci. Proc.* **12**, 1222-1234 (1991).
10. K.T. Lee and P.B. Aswath, *Mater. Sci. Eng. A* **352**, 1-7 (2003).
11. D. Long-González, J. López-Cuevas, C.A. Gutiérrez-Chavarría, P. Pena, C. Baudín, X. Turrillas, *Ceram. Int.* **36**, 661-672 (2010).
12. C.E. Semler and W.R. Foster, *J. Am. Ceram. Soc.* **52**, 679-680 (1969).
13. M. Oliveira and J.M.F. Ferreira, *Mater. Sci. Eng. A* **344**, 35-44 (2003).
14. L.B. Kong, Y.Z. Chen, T.S. Zhang, J. Ma, F. Boey and H. Huang, *Ceram. Int.* **30**, 1319-1323 (2004).
15. K.N. Lee, D.S. Fox, J.I. Eldridge, D. Zhu, R.C. Robinson, N.P. Bansal and R.A. Miller, *NASA/TM-2002-211372* (2002).
16. K.H. Schuller, *Brit. Ceram. Trans. J.* **64**, 103-117 (1964).
17. C. Baudín and M.P. Villar, *J. Am. Ceram. Soc.* **81**, 2741-2745 (1998).
18. M.J. Ribeiro, D.U. Tulyagavov, J.M. Ferreira and J.A. Labrincha, *J. Eur. Ceram. Soc.* **25**, 703-710 (2005).
19. S. Boŝković, D. Kosanović, Dj. Bahloul-Hourlier, P. Thomas and S.J. Kiss, *J. Alloy. Compd.* **290**, 230-235 (1999).

Mater. Res. Soc. Symp. Proc. Vol. 1373 © 2012 Materials Research Society
DOI: 10.1557/opl.2012.295

Physical, Petrographic, and Mineralogical Properties of Limestone Rocks from the Peninsula of Yucatán

J. May-Crespo[1], P. Quintana[1], J. J. Alvarado-Gil [1], B. A. Juárez de la Rosa[2], A. May-Pat[3], F. Avilés[3]

[1]Applied Physics Department, Cinvestav-Unidad Mérida, Carr. Ant. a Progreso km 6, C.P. 97310 Mérida, Yucatán, México.
[2]Marine Resources Department, Cinvestav-Unidad Mérida, Carr. Ant. a Progreso km 6, C.P. 97310 Mérida, Yucatán, México.
[3]Centro de Investigación Científica de Yucatán, Unidad de Materiales, Calle 43, No.130, Col. Chuburná de Hidalgo, CP 97200, Mérida, Yucatán, México.

ABSTRACT

Knowledge of the mechanical and petrographic properties of limestone rocks is an important issue to different areas of science and engineering. Sedimentary limestone rock is one of the most abundant materials in the Peninsula of Yucatán used for decorative and building construction. This work studies the petrographic, mineralogical, and physical properties of three different types of limestone slabs of the state of Yucatán.

INTRODUCTION

The study of properties of rocks has important applications [1]. Physical and chemical properties have always been considered important for several applications of the rocks. Two of the most commonly studied topics are mechanical properties and petrographic analysis [2]. Limestone is the most abundant type of rock in the state of Yucatán, where is an important building material, although it exhibits non-homogeneous physical and mechanical properties [3]. This material has been used by the Mayan civilization to construct their monumental buildings and roads that have lasted over time without important subsidence and deformation [4, 5]. Maldonado *et al.* [6] studied the weathering degradation rates of two types of limestone and proposed degradation kinetics mechanisms, although the mechanical properties of the analyzed limestone are not evaluated. Nowadays, limestone is used for various purposes in the building industry, for the renovation of historical monuments (e.g. stonework and sculptures). However, the mineralogical compositions, physical and mechanical properties of calcareous rocks are yet poorly characterized. In the present work, three different types of limestone rocks collected from different quarries of the state of Yucatán, México are characterized and their mechanical properties are determined.

EXPERIMENTAL

Selected materials

Three types of limestone rocks are collected from different quarries of the Peninsula of Yucatán, named as DA, DB and DR. DA and DB are yellow and white-colored rocks, respectively, and have low porosity, while DR is a compact red rock with higher porosity.

Characterization

Determination of physical properties such as density, water absorption, water content, and effective porosity, is carried out according to ASTM standard D2216-98 [7]. A total of 9 cubic samples of 1 cm^3 are prepared by using a precision cutter (Buehler Isomet), and the measurements are repeated 3 times for each rock. Megascopic petrographic description is made with a stereoscopic microscopy (Zeiss 47 50 2). Also, the analysis of 30 μm rock samples is carried out using polarized light (transmission and reflection) in an optical microscope (Karl Zeiss, Axiolab Pol) equipped with an image analyzer KS-300 and a petrographic microscope (Leica DM LP). X-ray diffraction (XRD) studies are carried out in a Siemens D-5000 diffractometer operated at 34 kV and 25 mA with CuKα radiation registered in the 3-45° (2θ) range, with a step size of 0.02° and step time of 12 s. Mechanical testing is performed according to ASTM standard D3148 [8]. Compression tests are conducted in a Shimadzu AGI-100 universal testing machine equipped with a 5 kN load cell and using a cross-head speed of 0.05 mm/min. A total of 21 samples are analyzed (7 of each rock), with average dimensions of 14.8 mm × 7.2 mm × 7.2 mm.

RESULTS

The measured physical properties of each rock are shown in Table 1. The density and water absorption values of samples DA and DB are similar (around 2.5 g/cm^3 and 1.22 %, respectively) and higher than those of DR (2.34 g/cm^3 and 0.98 %). Water content is similar for all samples (0.03 %) and the effective porosity is slightly higher for DA and DR (1.48% and 1.62%), than DB (1.16%).

Table 1. Physical properties of rock samples.

Samples	Density (g/cm^3)	Water absorption (%)	Water content (%)	Effective porosity (%)
DA	2.55	1.23	0.04	1.48
DB	2.51	1.22	0.03	1.16
DR	2.34	0.98	0.04	1.62

The optical microscope images and photomicrographs taken with parallel and polarized light for all samples are presented in Figure 1. Sample DA (Fig. 1a) shows the presence of microorganisms and marine fragments, which present a border of micritic material with the yellow color due to the low concentration of iron minerals. Rock DB (Fig. 1d) shows a uniform white matrix and a crystalline texture, with compact and defined particles. Sample DR (Fig. 1g) is very similar to DA. However, the hue is different between them due to the higher presence of iron oxides and hydroxides and shows a higher porosity.

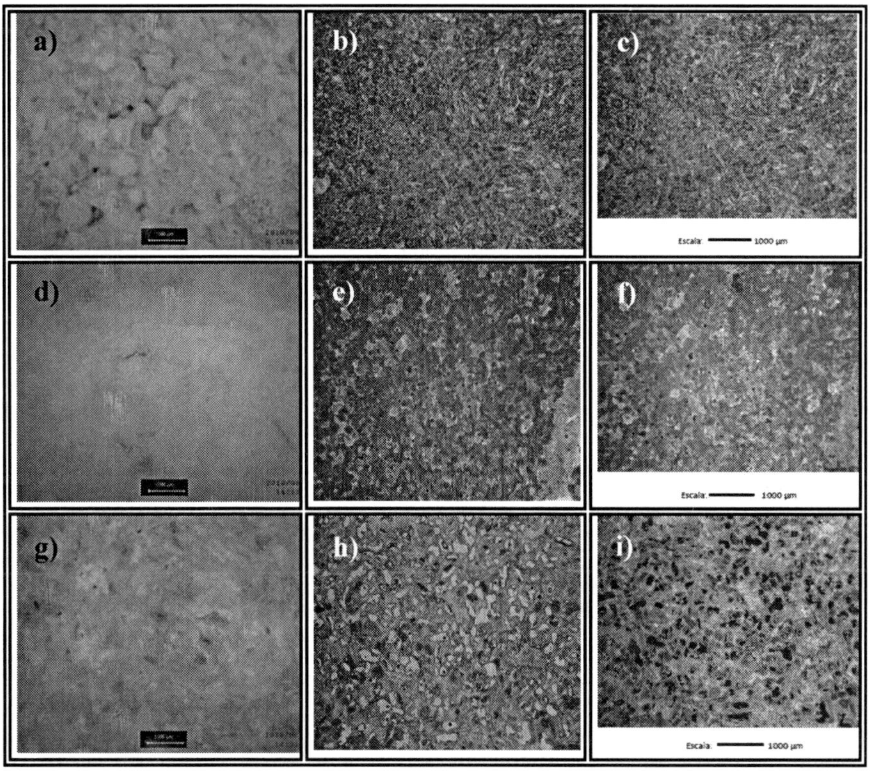

Figure 1. (a, d, and g) optical microscope images for the samples DA, DB and DR, respectively. (b, e, and h) photomicrographs taken with parallel light for DA, DB and DR, respectively. (c, f, and i) photomicrographs taken with polarized light for DA, DB and DR, respectively.

The petrographic description indicates that all three types of samples are carbonate rocks (Table 2), composed of micritic calcite for DA and DB, and sparitic calcite for DR. All samples presents microfossils, with a predominately moldic porosity with different percentages, DB exhibiting lower porosity than DA and DR. DA and DR samples are associated with iron oxides and hydroxides. Finally, the classification for DA and DB are biomicrite compacted and slightly cemented biomicrite, respectively, and DR sample is biosparite calcite [9].

XRD analysis of the rocks is performed on slices of 2 mm thickness. The predominant crystalline phase is calcite ($CaCO_3$). Traces of other crystalline phases are detected as dypingite ($Mg_5(CO_3)_4(OH)_2{\cdot}8H_2O$), quartz ($SiO_2$) and dolomite ($CaMg(CO_3)_2$). Starkeyite phase ($MgSO_4{\cdot}4H_2O$) is present only in sample DB. Finally, goethite ($FeO(OH)$) is detected on samples DA and DR.

Table 2. Petrographic description and classification of studied rocks.

Samples	Petrographic description	Classification +Folk,1962 ++Wright,1992
DA	Carbonated rock with a pink earthy-looking brown color composed of micritic calcite with associated particles of Fe oxides and hydroxides. Microfossils consist of several foraminifers. Ooids are subrounded particles covered with a layer of micrite and Fe oxides and hydroxides. Porosity is high dominated by moldic pores.	+Biomicrite compacted ++Grainstone
DB	Carbonated white color rock, composed of calcite and sparite intraparticles. Microfossils are formed of foraminifers. Porosity is low, predominately moldic.	+Slightly cemented Biomicrite ++Grainstone
DR	Carbonated rocks with an ocher with mottled brown color, consisting of calcite sparite associated with iron oxides and hydroxides. Microfossils are made of foraminifers. Ooids are seen as round particles. Porosity is high and is predominantly moldic.	+Biosparite ++Cemented rock

The percentage of calcium carbonate is very high in all samples (98%). Therefore, in order to identify clay minerals in each sample, they are subjected to dissolution of carbonates with HCl 5% v/v. XRD analysis of the powders after the dissolution (Fig. 2) shows that the mineral phase composition is very similar for sample DA and DR. The presence of quartz, goethite, tosudite ($(KCa)O_8Al_6(SiAl)_8O_{20}(OH)_{10}4H_2O$), and hematite ($Fe_2O_3$) are detected. On the other hand, anatase (TiO_2), montmorillonite ($Ca_{0.2}(Al,Mg)Si_4O_{10}(OH)_24H_2O$) and kaolinite ($Al_2Si_2O_5(OH)_4$) are identified in sample DB.

The results of the mechanical properties obtained from compressive testing are shown in Figure 3. The stress-strain (σ-ε) curves for samples DA and DB presented a similar behavior, since their physical properties are similar as seen in Table 1. The reduced porosity of DB with respect to DA may cause that the initial elastic slope is smaller for DB. They have an average maximum strength around 90 MPa and ultimate strain of 1.47 % and 1.21% respectively (see Table 3). On the other hand, sample DR has a very different mechanical behavior with reduced strength and ultimate strain, evidencing a more brittle behavior, which is in direct correspondence with its increased porosity. Regarding the elastic modulus (E), samples DA and DB have an E of around 8.20 GPa while the modulus of DR is slightly lower (7.5 GPa). Notice that the elastic modulus keeps a correlation with the sample density listed in Table 1. According to the obtained strength (σ_{max}) of the tested rocks, samples DA and DB are classified as rocks with a high resistance to compression, while DR is a medium resistance rock [10].

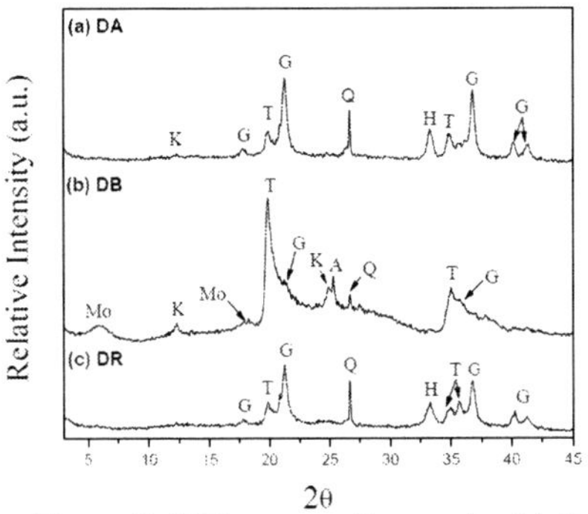

Figure 2. XRD patterns for samples (a) DA, (b) DB, and (c) DR, respectively, after the dissolution of calcium carbonate. Where: K=Kaolinite, G=Goethite, T=Tosudite, Q=Quartz, A=Anatase, Mo=Montmorillonite, H=Hematite.

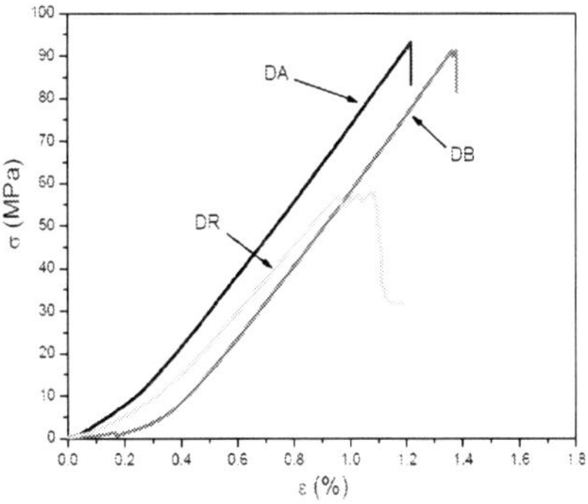

Figure 3. Stress-strain curves obtained by compression tests.

Previously reported compressive test experiments by Alonzo [1] showed that limestone from Yucatán has an average strength of 28.6 MPa, with maximum values of 64.7 MPa. Later, others studies made by Estrada [5] showed that a rock similar to DR exhibited strength and ultimate strain values of 18.5 MPa and 1%, respectively. However, the mineralogical and petrographic properties are not evaluated in those works [1, 5, 6]. In comparison with the results obtained in this work, the average strength and strain values are higher due to the mineralogical composition and the low porosity. Another studies on limestone rocks from Turkey reported a similar compressive strength range of 93-126 MPa [11, 12] for rich calcite (95%) rocks classified as biomicrite [11]. Other investigations on rich dolomite limestone rocks [2] resulted in a very low compressive strength of 16 MPa. These differences in mechanical properties are due to the differences in microstructural homogeneity, porosity and composition exhibited by each type of limestone rock investigated in each work.

Table 3. Mechanical properties of rock samples.

Samples	σ_{max} (MPa)	ε_{max} (%)	E(GPa)
DA	90.8 ± 2.86	1.21 ± 0.01	8.13 ± 0.21
DB	91.4 ± 2.45	1.47 ± 0.16	8.34 ± 0.22
DR	55.8 ± 2.95	0.95 ± 0.01	7.47 ± 0.37

CONCLUSIONS

Laboratory test of the mineralogical, and physical properties of different types of limestone rocks obtained from the state of Yucatán, México, are carried out and the relationships among their properties are discussed. The density and water absorption values for samples DA and DB are similar; however effective porosity is lower for DB. Rocks DA and DB are classified as compacted biomicrite and slightly cemented biomicrite, respectively, and DR is biosparite. This is because rocks DA and DB are constituted from micritic calcite, while DR from sparitic calcite. All the studied samples are mainly composed of calcite and have minor amounts of quartz, goethite, anatase and tosudite, besides DA and DR show the presence of hematite and DB has a minor amount of montmorillonite. A similar mechanical behavior is obtained for samples DA and DB, with high values of strength, ultimate strain, and modulus of elasticity. Sample DR, however, has a reduced strength and ductility in comparison with DA and DB, since its effective porosity is higher which yields a lower density.

ACKNOWLEDGEMENTS

J. May-Crespo acknowledge at CONACyT (M.Sc. grant-172739), and for the financial support of projects 123913 and FOMIX 108528, 108160.

REFERENCES

[1] L. A. Alonzo-Salomón, L. Espinoza-Graham, *Ingeniería*, **7-1**, 27-36, (2003).
[2] E.A. Ozsoy, G. Yilmaz, H. Arman, *Sci. Res. Essays*, Vol. **5** (17), 2579-2587, (2010).
[3] L. Espinoza, M. Cerón, A.Sulub, *Int. J. Rock Mech. Min. Sci.*, Vol. **35**, No.4/5, 410-411, (1998).
[4] J. May-Crespo, P. Martinez-Torres, J.J. Alvarado-Gil, P.Quintana, J. Ordoñez-Miranda, *Int. J. Thermophys,* **31**, 1027–1036, (2010).
[5] H. Estrada-Medina, S. Valdez, A. Zanatta, S.R. Casolco. Memorias del 14 Congreso Internacional Anual de SOMIM, (2008).
[6] L. Maldonado, L. Veleva, L. Díaz-Ballote**,** *Appl. Phys. A,* **103**, 1105–1110, (2011).
[7] ASTM D2216-98. Standard test method for laboratory determination of water (Moisture) content of soil and rock by mass. West Conshohocken (PA, USA): ASTM international; 2007.
[8] ASTM D3148. Standard test for elastic moduli of intact rock core specimens in uniaxial compresion. West Conshohocken (PA, USA): ASTM international; 2007.
[9] R. Flok, *Petrology of sedimentary rocks*, Hemphill Publishing Company. Austin, Texas, (1980).
[10] Manual de diseño de obras civiles, B.3.4, Comisión Federal de Electricidad, México, (1979).
[11] A. Tugrul. *Eng. Geol.*, **75**, 215-227, (2004)
[12] H. Yavuz, R. Altindag, S. Sarac, I. Ugur, N. Sengun, *Int. J. Rock Mech. Min. Sci.,* **43**, 767-775, (2006).

Mater. Res. Soc. Symp. Proc. Vol. 1373 © 2012 Materials Research Society
DOI: 10.1557/opl.2012.296

Effect of Gradual Substitution of CaO by SrO in Glass-Ceramic Materials of the System SiO_2 - Al_2O_3 - CaF_2 - RO (R = Ca, Mg, Sr)

M. Garza-García[1], J. López-Cuevas[2], C.A. Gutiérrez-Chavarría[2], N. Piedad-Sánchez[3], E. Camporredondo-Saucedo[3], O. Hernández-Ibarra[4]
[1]Universidad Autónoma de Coahuila, Escuela Superior de Ingeniería, Nueva Rosita, Coahuila, México.
[2]Centro de Investigación y Estudios Avanzados del IPN, Unidad Saltillo, Coahuila, México.
[3]Facultad de Metalurgia. Universidad Autónoma de Coahuila, Monclova, Coahuila, México.
[4]Instituto Tecnológico de Estudios Superiores de la Región Carbonífera, Coahuila, México.

ABSTRACT

The density, Vickers microhardness and crystallization fraction of glass-ceramic materials synthesized from parent glasses are determined in which CaO is gradually substituted by SrO. The chemical composition (in mol.%) of the parent glasses is $54SiO_2$-(23-X)CaO-12MgO-$5Al_2O_3$-$6CaF_2$-XSrO, where X is the employed CaO substitution level (X = 0, 3, 6 and 9 mol.%, with X = 0 corresponding to the reference material). In order to determine the type of crystallization occurring in the glass-ceramic samples, as well as the crystalline phases formed in them, these are characterized by both Scanning Electron Microscopy (SEM/EDS) and X-Ray Diffraction (XRD). Independently of the CaO substitution level employed, the glass-ceramics show the formation of a solid solution corresponding to diopside-type pyroxene, with chemical formula $Ca(Mg,Al)(Al,Si)_2O_6$, as a single crystalline phase. The synthesized glass-ceramic materials with the reference composition show the highest Vickers microhardness and crystallization fraction, as well as the lowest density.

INTRODUCTION

Glass-ceramic materials are obtained through the controlled crystallization of a parent glass, involving the separation of a crystalline phase from the glass. The final properties of these materials are determined by the size, morphology, composition, properties and distribution of the crystals dispersed in the glassy phase [1].

Diopside ($CaMgSi_2O_6$) is a pyroxene mineral belonging to the single chain group of the inosilicate family, with a simple crystalline structure in which the SiO_4 tetrahedra are linked to each other by two oxygen atoms forming a relatively seemingly endless chain. The basic structural group is Si_2O_6 with silicon to oxygen ratio of 1:3. The most common crystalline morphologies presented by diopside are needle and dendritic. Diopside is also characterized by high twinning, which constitutes an energy-absorbing mechanism. Thus, a glass-ceramic material is strengthened, getting high hardness and resistance to fracture, if it has diopside within its crystal structure [2]. On the other hand, it is known that SrO improves some properties of glasses, such as their hardness and abrasion resistance. SrO also increases the brightness and ease of polishing of glasses and it is used too in the glazing of traditional ceramics, preventing the formation of bubbles during the firing process of the frit [3]. Thus, a glass-ceramic material having diopside as main crystalline phase, and containing SrO in the remaining glassy phase, has a potential to possess interesting properties, which can make it suitable for structural applications, for instance in the form of tiles, dental and biomedical components, refractories, etc. [4-8].

This research work is aimed to develop glass-ceramic materials with a base composition in the primary crystallization field of diopside [$CaMgSi_2O_6$] in the quaternary system SiO_2-CaO-MgO-Al_2O_3. The effect of gradual substitution of CaO by SrO on the crystallization of the parent glasses and on some properties of the synthesized glass-ceramics is studied.

EXPERIMENTAL DETAILS

The raw materials are: silica sand (purity of 99.98wt%, 110μm, MERASI, México); Al_2O_3 (HPA-0.5, 99.99 wt%, 0.75μm, Sasol, EUA); dolomite ($(Ca,Mg)(CO_3)_2$, with ~3wt% of $CaCO_3$ as main impurity, 35μm, Química del Rey, México); calcite ($CaCO_3$, 99.6 wt%, 45μm, Materiales La Gloria, México); fluorite (CaF_2, with 10.3wt% SiO_2, 2.2wt% Al_2O_3 and 19.8wt% $CaCO_3$ as main impurities, 30μm, Materiales La Gloria, México), and $SrCO_3$ (97.86wt%, 4μm, Solvay, México). For the synthesis of the parent glasses, four mixtures are first prepared using suitable proportions of the raw materials, according to the intended levels of CaO substitution. All initial mixtures are homogenized for 1h in a ball mill and melted at 1450°C/3h in a Pt crucible using a Lindberg/blue M BF51433 PC-1 high temperature electric furnace. The melt is cast on a steel plate previously heated to 600°C, forming monoliths of several sizes (Figure 1). The glassy monoliths are subsequently subjected to an annealing treatment in a Thermolyne F62700 electric furnace, in which they are cooled down from 600°C to room temperature at a rate of 5°C /min during the first hour and at a rate of 1°C/min during the following nine hours. After that the glass-ceramics are obtained by crystallization of the glass monoliths at 1000°C/3h, using a Thermolyne 46200 high temperature electric furnace, with heating rate of 5°C/min and cooling rate of 10°C/min.

For microstructural analysis, the synthesized glass-ceramics are mounted in cold-cure epoxy resin and then ground using SiC papers with successive grit sizes from 80 grit to 2400 grit. This is followed by polishing to a mirror finish using diamond particles with sizes of 3, 1 and ¼ μm, and then by graphite coating by using a JEOL JEE-400 vacuum evaporator. The analysis of the samples is carried out by Scanning Electron Microscopy (SEM/EDS) using a Phillips XL30 ESEM, with an accelerating voltage of 20 KV and a working distance of 10 mm.

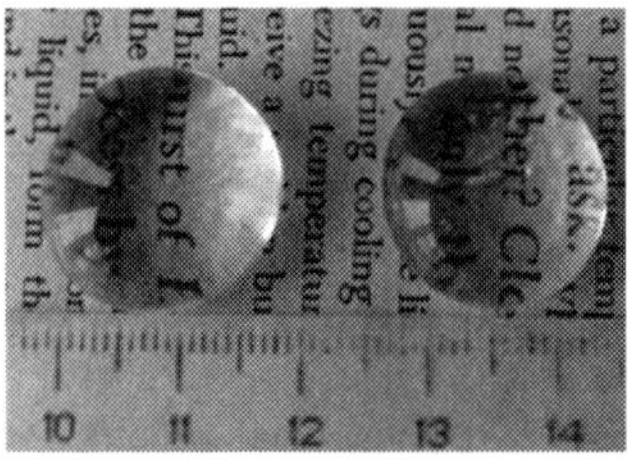

Figure 1. Glass monoliths.

X-Ray Diffraction (XRD) is used for the identification of the crystalline phases formed in the synthesized glass-ceramics as well as for the estimation of their degree of crystallinity. These analyses are carried out by using a Philips X'Pert 3040 apparatus, CuKα radiation (λ = 1.542 Å), an accelerating voltage of 40 kV and a current of 30 mA, in the 2θ range of 10–80°, with a step size of 0.035° (2θ/s). The degree of crystallinity, X(%), is estimated by using a method developed by Ohlberg and Strickler [9], which employs equation (1):

$$X(\%) = \frac{I_g - I_x}{I_g} \times 100 \qquad (1)$$

I_g and I_x correspond to the integrated intensities of the background lines of the XRD patterns obtained for the parent glass and for a glass-ceramic sample, respectively.

Prior to the measurement of the density of a glass-ceramic material by the Archimedes' principle, using a Quantachrome MVP-D150E helium pycnometer, a small portion of it is milled by using agate mortar and pestle until a particle size corresponding to the 100 mesh sieve (150 μm) is obtained. Then, the density of the powdered material is measured three times, reporting the average value for each one of the studied compositions.

Vickers microhardness (HV) testing of the glass-ceramic materials is carried out by using a Wilson Tukon 300-FM apparatus, applying a load of 500 gf during 5s. The measurements are made in ten randomly selected points for each one of the samples. Since in all cases the indentation size is big enough, it is considered that a mean HV value for all crystalline and amorphous phases present in the microstructure of the material is obtained. The measurement of the Vickers microhardness is based on the ASTM E384-99 standard.

DISCUSSION

Phase composition of synthesized glass-ceramics

All samples show the formation of a crystallized shell surrounding a glassy core (Figure2a). In all cases, the crystallized surface layer had a thickness of ~750 μm and a microstructure consisting of feather-like dendritic crystals dispersed in a glassy matrix (Figure 2b). As reported in the literature [2], this type of crystal growth is characteristic of pyroxenes. In this case, diopside-type pyroxene is formed as a single crystalline phase, according to the results obtained from the XRD analyses that are carried out. Figure 3 shows the XRD patterns obtained for samples with CaO substitution levels of X = 3, 6 and 9.

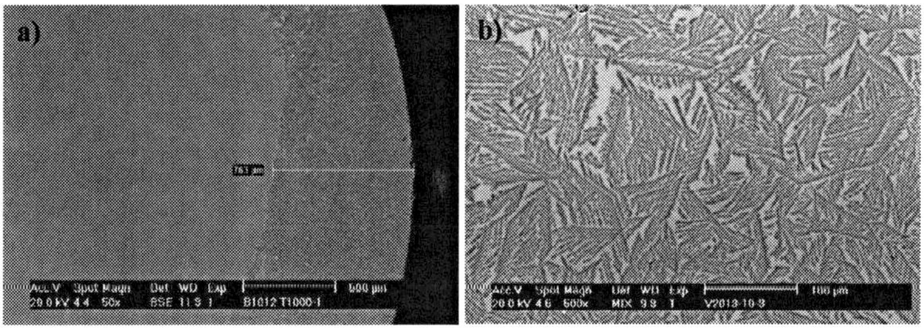

Figure 2. SEM micrographs of heat-treated samples: a) crystallized surface layer formed in a sample with X = 6, and b) microstructure of the crystallized surface layer formed in a sample with X = 9; feather-like dendritic crystals correspond to diopside-type pyroxene phase dispersed in a glassy matrix.

During the formation of the diopside-type pyroxene phase, the incorporation in it of different oxides in solid solution slightly modified its lattice parameters with respect to those corresponding to pure diopside. This is evidenced by a slight shift in the position of the corresponding XRD peaks in the heat-treated samples towards lower or higher 2θ angles when compared with the XRD pattern of pure diopside [5]. This is observed in all our samples. An example of this can be seen in Figure 4. Thus, it is determined that the diopside-type pyroxene phase formed in our samples has a chemical formula $Ca(Mg,Al)(Al,Si)_2O_6$, instead of the chemical formula $CaMgSi_2O_6$ corresponding to pure diopside. This is corroborated by the SEM/EDS analysis that is carried out.

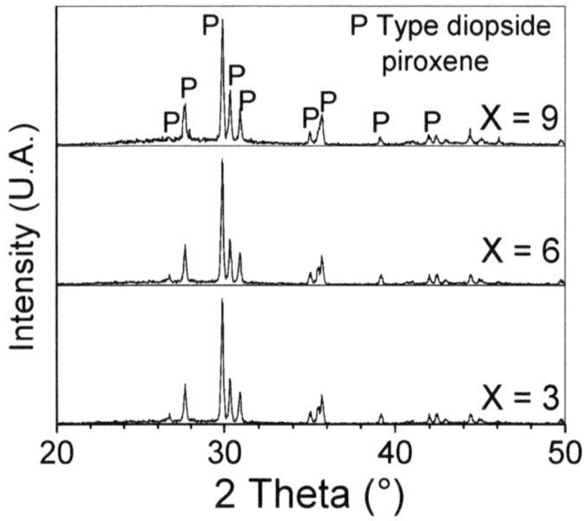

Figure 3. XRD patterns obtained for heat-treated samples with CaO substitution levels of X = 3, 6 and 9. P = diopside-type pyroxene phase.

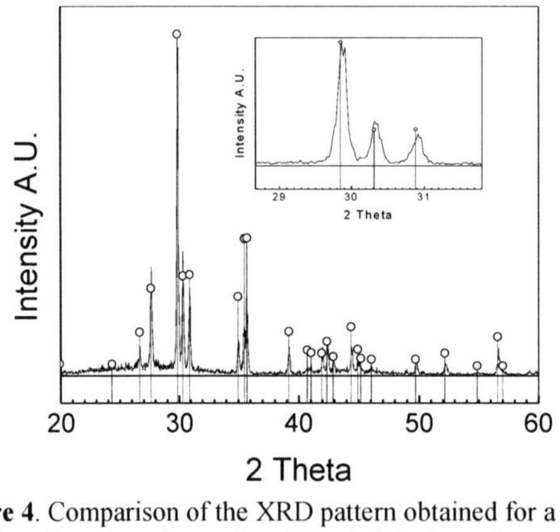

Figure 4. Comparison of the XRD pattern obtained for a sample with a CaO substitution level of X = 6, against that of pure diopside (o). The inset shows more clearly the peak shift.

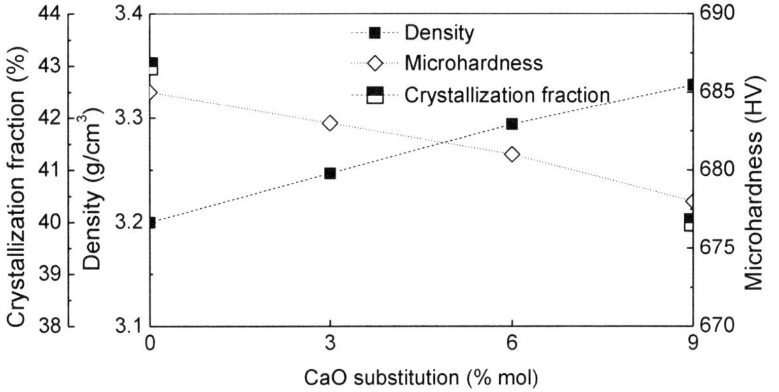

Figure 5. Effect of gradual substitution of CaO by SrO on the density, crystallization fraction and Vickers microhardness (HV) of the synthesized glass-ceramics.

Properties of synthesized glass-ceramics

Figure 5 shows the effect of gradual substitution of CaO by SrO on the density, Vickers microhardness and crystallization fraction of the heat-treated samples. As can be seen, the density of the glass-ceramics increases with increasing CaO substitution level. The opposite effect is caused on their crystallization fraction and Vickers microhardness. This can be due to a combined effect of the following factors: a) the substitution of Ca^{2+} by other cations with lower field strength, such as Sr^{2+} causes softening of the residual glassy phase in the glass-ceramic, generating a net diminution in the microhardness of the latter materials, b) density increases due to the substitution of Ca^{2+} by other cations with higher atomic weight, such as Sr^{2+}, and c) due to their relatively larger size, the Sr^{2+} cations can block the passage of the smaller Ca^{2+} and Mg^{2+} cations required to form the diopside phase, causing a delay in the crystallization process.

CONCLUSIONS

Crystallization begun at the surface of the glass samples and the crystals grew toward the central zone of the monolith, where a core of the original parent glass remained. The density of the glass-ceramics increased, while their crystallization fraction and Vickers microhardness decreased, with increasing level of substitution of CaO by SrO in the parent glass. This is attributed to differences existing between the Ca^{2+} and Sr^{2+} cations regarding their field strengths, atomic weights and ion sizes. Crystals of a solid solution corresponding to diopside-type pyroxene phase, $Ca(Mg,Al)(Al,Si)_2O_6$, are formed in all glass-ceramic samples. The SEM/EDS analysis corroborated this. Lastly, although diopside-containing glass-ceramic materials can be produced by partly replacing CaO with SrO in the parent glass, it is found that among the developed materials the one with no replacement of CaO (X = 0) show the highest Vickers microhardness, crystallization fraction as well as the lowest density.

ACKNOWLEDGMENTS

The authors express their gratitude to CONACYT, U.A. de C. and Cinvestav-Saltillo for the financial support and facilities provided for the development of this work.

REFERENCES

1. J. Doyle, *Glass Making Today*, 242-243 (1979).
2. W. Höland and G. Beall, *Glass-Ceramic Technology*, (The American Ceramic Society Westerville, Ohio, 2002).
3. www.qsrestroncio.com
4. C. Leonelli, T. Manfredini, M. Paganelli, P. Pozzi and G.C. Pellacani, *J. Mat. Sci.* **26**, 5041-5046 (1991).
5. F.J. Torres and J. Alarcón, *J. Eur. Ceram. Soc.* **25**, 349–355 (2005).
6. S.A.M. Abdel-Hameed and A.A. El-kheshen, *Ceram. Int.* ***29***, 265–269 (2003).
7. S.N. Salama, H. Darwish and H.A. Abo-Mosallam, *J. Eur. Ceram. Soc.* **25**, 1133–1142 (2005).
8. D.U. Tulyaganov, S. Agathopoulos, H.R. Fernandez, J.M. Ventura and J.M.F. Ferreira, *J. Eur. Ceram. Soc.* ***24***, 3521–3528 (2004).
9. S.M. Ohlberg and D.W. Strickler, *J. Am. Ceram. Soc.* **45**, 170 (1962).

Mater. Res. Soc. Symp. Proc. Vol. 1373 © 2012 Materials Research Society
DOI: 10.1557/opl.2012.297

Distribution of Additives in Single-Modified TiO_2

Marta A. Gleń[1] and Barbara U. Grzmil[1]
[1]West Pomeranian University of Technology, Szczecin, Institute of Chemical and Environment engineering, 70-322 Szczecin, Pułaskiego 10, E-mail: mglen@zut.edu.pl

ABSTRACT

In the present work the influence of different modifiers, calculated to B_2O_3, CeO_2, Sb_2O_3, ZnO, and ZrO_2, on their distribution in TiO_2 is investigated. The phase composition and phase transformation of prepared rutile-TiO_2 is determined by the selective leaching method, ICP-AES, XRD and FT-IR techniques. The addition of Sb_2O_3 to TiO_2 has no influence on the anatase–rutile phase transformation, CeO_2 and ZrO_2 act as inhibitors of the TiO_2 phase transformation and the addition of ZnO or B_2O_3 to TiO_2 accelerates rutile formation. It is observed that boron is located in TiO_2 in the form of soluble B_2O_3, zinc partly reacts with titanium forming co-phase $TiZn_2O_4$ and antimony addition to TiO_2 presumably causing the formation of a co-phase of Sb with Ti. Cerium forms a separate phase, CeO_2, and reacts partly with titanium, probably creating co-phase, $Ce_xTi_{(1-x)}O_2$ (for example $Ce_{0.8}Ti_{0.2}O_2$). Zirconium addition in TiO_2 forms separate ZrO_2 phase and solid solution of Zr with Ti.

INTRODUCTION

Titanium dioxide crystallizes in three naturally occurring polymorphous modifications: anatase, brookite and rutile. Anatase is a superior photocatalytic material for air and water purification, water disinfection and hazardous waste remediation [1]. Rutile plays the most important role in almost all inorganic pigments. Rutile pigment is mainly used in paint and coating industries, but additionally it is widely applied in paper, artificial fibre, plastic and rubber sectors of the market [2].

The polymorphous anatase-rutile phase transformation is affected by the kind and contents of additives present in calcined titanium dioxide. Their distribution both on the surface and in the bulk of titanium dioxide can act as promoters or inhibitors of the transformation, presence of rutile nuclei, particle size, temperature and calcination time [3-7].

EXPERIMENT

The starting material is technical grade hydrated titanium dioxide (HTD) as a concentrated suspension containing 40 wt % of TiO_2. It is an indirect product from the industrial production of TiO_2 with the sulfate method. The modifier solutions are prepared by dissolving Sb_2O_3 in HCl aqueous solution and H_3BO_3, $Ce(NO_3)_3$, $ZnSO_4$, $Zr(SO_4)_2$ in distilled water. Solutions of modifying agents (calculated to B_2O_3, CeO_2, Sb_2O_3, ZnO, ZrO_2) are introduced to HTD. Contents of B_2O_3, CeO_2, Sb_2O_3, ZnO, or ZrO_2 in TiO_2 are changed respectively: 0.06-0.34 mol%, 0.02-0.14 mol%, 0.01-0.08 mol%, 0.05-0.3 mol%, and 0.03-0.19 mol%.

The obtained pulp after thorough mixing is transferred to an evaporating dish and inserted into a laboratory muffle furnace heated to the desired temperature. The prepared samples, for investigations of the anatase-rutile phase transformation and modifiers' distribution, are calcined for 1 h at 700, 750, and 800 °C for TiO_2 modified with B_2O_3, Sb_2O_3, or ZnO and for 1 h at 700,

750, 800 and 850 °C for TiO_2 modified with CeO_2 or ZrO_2. Unmodified TiO_2 is calcined at 700-850 °C. TiO_2 samples modified with 4,6 mol% B_2O_3, 0.93 mol% CeO_2, 0.53 mol% Sb_2O_3, 2 mol% ZnO, or 2-7.8 mol% ZrO_2 for the purpose of X-ray diffraction analysis are calcined for 2 h at 1000°C. Additionally, pure B_2O_3 for FT-IR measurements is obtained at 1000 °C.

The X-ray diffraction analysis (X'Pert PRO Philips diffractometer with CuK_α radiation) is used to determine the content of rutile and phase composition in the calcined titanium dioxide samples. The relative abundance of anatase and rutile phases is calculated from the (101) reflection of anatase and the (110) reflection of rutile. The content of rutile in the samples is determined from the following equation:

$$W_R = 100/(1+I_A/k \cdot I_R) \quad (1)$$

where: W_R is the content of rutile, I_A and I_R are the peak intensities of anatase (101) and rutile (110) and k is the coefficient (the ratio of peak intensity (101) 100 wt % of anatase to the peak intensity (110) 100 wt % of rutile).

FT-IR measurements (FT/IR-430 Jasco spectrophotometer, wavenumber range 4000 to 400 cm^{-1}) are conducted for investigations of TiO_2 phase composition.

Distribution of modifiers in the investigated samples of TiO_2 is determined with a selective leaching method [8]. Solution of ammonia (0.5 mol/L), sodium-EDTA (0.1 mol/L) and hydrochloric acid (6 mol/L) are used for leaching the modifiers and titanium from the calcined samples of TiO_2. After calcination, TiO_2 samples are milled and mixed with an appropriate leaching solution. The prepared suspension is shaken in a water bath shaker at room temperature for 2h. The solid phase is separated by centrifuging (High Speed Brushless Centrifuge MPW-350, 8000 rpm, 20 min). The contents of elements and Ti in the liquid phase are determined with the ICP-AES method (Optima 5300 DV, Perkin Elmer). The leaching degree is defined as a ratio of the weight of the element leached into the liquid phase to the weight of that element contained in the titanium dioxide before leaching.

DISCUSSION

The content of the rutile phase in the calcined TiO_2 samples is determined by X-ray diffraction analysis in order to determine the effect of particular modifiers on the anatase–rutile phase transformation. Based on the obtained XRD patterns, the transformation degrees are calculated according to Eq. (1). Referring to the data (Fig. 1), it is clearly shown that the Sb_2O_3 addition to the TiO_2 samples has no influence on the anatase–rutile phase transformation, but the CeO_2 or ZrO_2 addition causes an increase in the rutilization temperature thus acting as inhibitors to the TiO_2 phase transformation. The addition of ZnO or B_2O_3 to TiO_2 accelerates rutile formation. This effect is observed only at intermediate calcination temperatures.

Differences in the effect of modifiers on the anatase-rutile phase transformation can result from different mechanisms of the reaction with TiO_2. In order to determine the phase composition of the TiO_2 samples, a selective leaching method is used. Characteristics of the unmodified and modified rutile TiO_2 and the results from the leaching of these materials with various solutions are shown in Table 1. The leached TiO_2 modified with B_2O_3, Sb_2O_3, or ZnO is calcined at 800 °C and modified with CeO_2 or ZrO_2 at 850 °C.

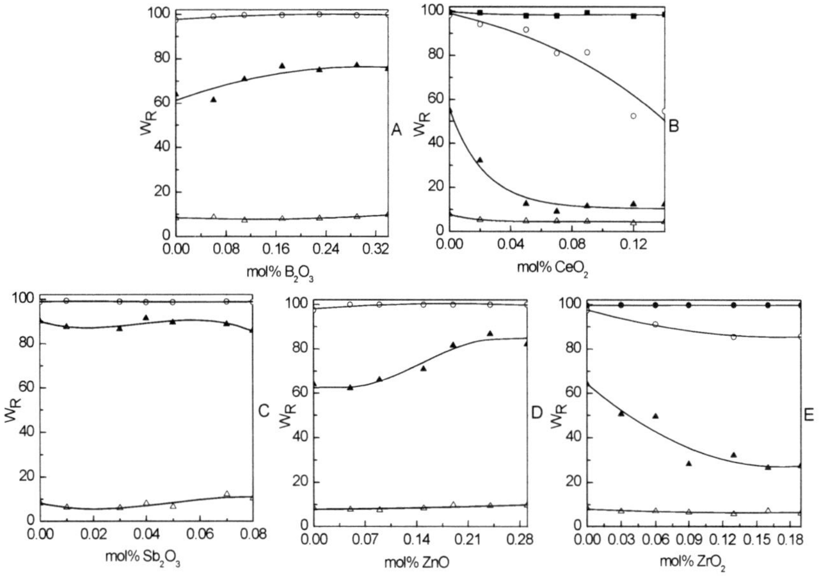

Figure 1. The influence of B_2O_3 (A), CeO_2 (B), Sb_2O_3 (C), ZnO (D), or ZrO_2 (E) modifier and calcination temperature on the anatase–rutile phase transformation degree (W_R); (△) 700 °C, (▲) 750 °C, (○) 800 °C, (●) 850°C.

The degree of boron leaching from modified TiO_2 amounted to about 100% for all the solutions used. This proves that boron is located in TiO_2 in the form of soluble compound, presumably B_2O_3. The XRD analysis could not confirm any formation of boron phase, probably because of its amorphous state. Therefore, FT-IR measurements are conducted additionally (Fig. 2). The band at 1200 cm^{-1} is assigned to the B-O stretching vibrations of BO_3 units and appeared in TiO_2 modified with 4.6 mol% B_2O_3. Moreover another band has been observed at 2255 cm^{-1}. In unmodified TiO_2 it does not exist. However, in comparison with obtained B_2O_3, an absorption at 2255 cm^{-1}also appears. That is to say TiO_2 is modified with B_2O_3.

The leaching degree of Ce in HCl and EDTA is 22 and 14%, respectively In the EDTA solution, no Ti is found whereas, in the solution of HCl, the content of Ti is 0.35 mmol/L. This attests that the Ce reacted with TiO_2 and formed a partially soluble phase. Separate CeO_2 phase, partially soluble in EDTA and HCl is also formed. The XRD pattern of TiO_2 modified with 0.14 mol% CeO_2 showed only the rutile phase. Therefore, TiO_2 modified with a higher amount of CeO_2 is prepared and investigated. The objective is to verify the phase formation It can be seen from Fig. 3 that for 0.93 mol% CeO_2, diffraction peaks at 2θ = 28.6°, 33.1°, 47.5°, 59.1°, and 79.1° can be attributed to both CeO_2 (JCPDS-04-008-4953) and $Ce_{0.8}Ti_{0.2}O_2$ (JCPDS-04-005-9314) phases.

Table 1. Contents of elements in different solutions after leaching of TiO_2 samples.

Content of additive type (mol%)	Solvent	Concentration (mol/L)	Content in solution		Leaching degree	
			Element (mmol/L)	Ti (mmol/L)	Element (%)	Ti (%)
0	HCl	6	-	0.15	-	0.02
	EDTA	0.1	-	-	-	-
	NH_4OH	0.5	-	0.1	-	0.02
0.34 B_2O_3	HCl	6	2.03	0.12	96	0.02
	EDTA	0.1	2.13	-	98	-
	NH_4OH	0.5	2.16	-	100	-
0.14 CeO_2	HCl	6	0.19	0.35	22	0.06
	EDTA	0.1	0.12	-	14	-
	NH_4OH	0.5	-	0.08	-	0.01
0.08 Sb_2O_3	HCl	6	-	0.15	-	0.02
	EDTA	0.1	-	-	-	-
	NH_4OH	0.5	-	0.15	-	0.02
0.29 ZnO	HCl	6	0.91	0.15	49	0.02
	EDTA	0.1	0.80	-	43	-
	NH_4OH	0.5	0.02	-	1	-
0.19 ZrO_2	HCl	6	-	0.12	-	0.02
	EDTA	0.1	-	-	-	-
	NH_4OH	0.5	-	-	-	-

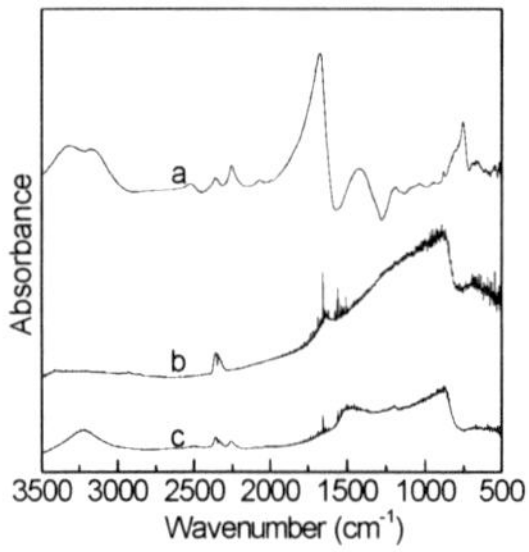

Figure 2. FT-IR spectra of samples calcined at 1000 °C; a – pure B_2O_3; b – unmodified TiO_2; c – TiO_2 modified with 4.6 mol% B_2O_3.

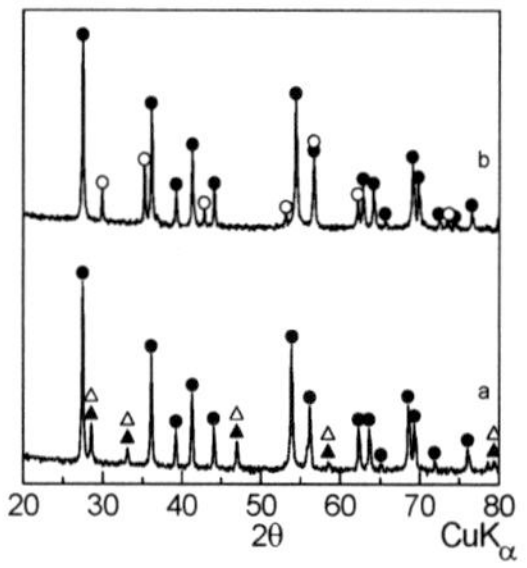

Figure 3. X-ray diffraction patterns of TiO_2 modified with 0.93 mol% CeO_2 (a) and 2 mol% ZnO (b); rutile (●), CeO_2 (△), $Ce_{0.8}Ti_{0.2}O_2$ (▲), $TiZn_2O_4$ (○).

In the solutions investigated, not even trace amounts of Sb are determined. The absence of Sb in HCl solution excluded the formation of any antimony oxides as long as Sb_2O_3, Sb_2O_4, and Sb_2O_5 are soluble in HCl. Only trace amounts of Ti are released into the solutions of HCl and NH_4OH. However, the same amounts of Ti are released into the same solutions during the leaching of unmodified TiO_2. This attests that Sb formed a co-phase with TiO_2. The XRD analysis could not confirm any formation of the antimony phase, even when present in higher content expressed as 0.53 mol% Sb_2O_3.

Zinc contents after leaching of modified TiO_2 in solution of HCl and EDTA is similar and amounted to 0.91 and 0.80 mmol/L, respectively. In EDTA solution no Ti is found whereas in solution of HCl the content of Ti is 0.15 mmol/L. It testifies that during the calcination process about 57% of zinc reacts with Ti and forms co-phase, $TiZn_2O_4$ which is slightly soluble in solution of HCl. Hence 43% of zinc remains as ZnO. The formation of co-phase with titanium and zinc, for 2 mol% ZnO in TiO_2, is shown in Fig. 3. Diffraction peaks at 2θ=29.9, 35.2, 42.8, 53.1, 56.7, 62.1 and 73.7° are characteristic of $TiZn_2O_4$ (JCPDS-01-077-0014). The evidence of solubility of this phase can be seen in Fig. 4. The TiO_2 sample modified with 2 mol% ZnO is also leached and subjected to XRD analysis. The decrease of $TiZn_2O_4$ peak intensity at 2θ=35.2° is observed after leaching with HCl solution.

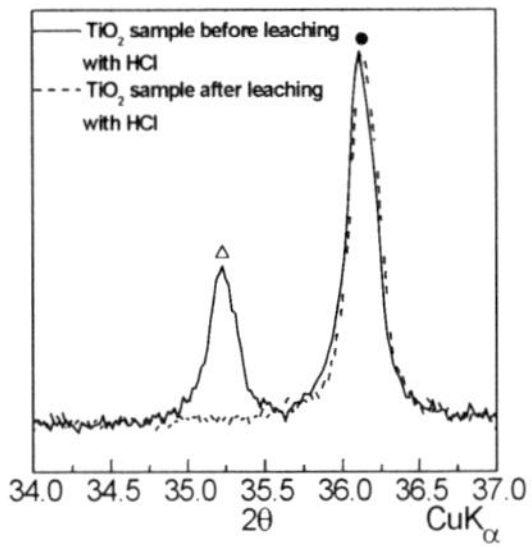

Figure 4. X-ray diffraction patterns of TiO_2 modified with ZnO calcined at 1000 °C before and after leaching with 6 mol/L HCl; rutile (●), $TiZn_2O_4$ (△)

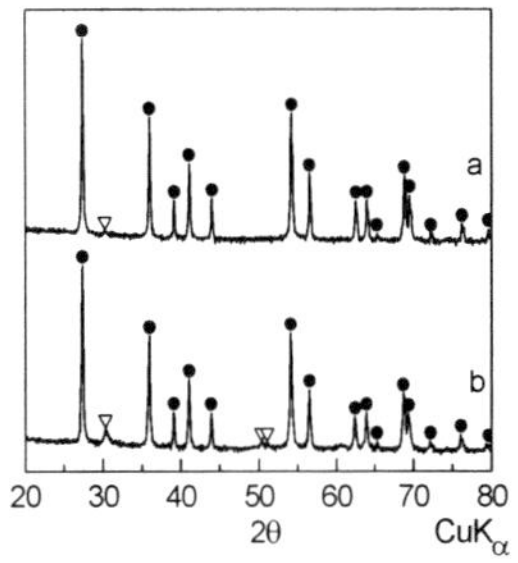

Figure 5. X-ray diffraction patterns of TiO_2 modified with ZrO_2 calcined at 1000 °C; rutile (●), ZrO_2 (▽); a – 4 mol% ZrO_2; b – 7.8 mol% ZrO_2

TiO_2 modified with Zr is leached. In obtained solutions even trace amounts of Zr are not determined. Only trace amounts of Ti are released to the solution of HCl. However the same amount of Ti is released to solution of HCl during the leaching of unmodified TiO_2. It testifies that Zr reacts with TiO_2 and forms partly soluble phase of a $Ti_{1-x}Zr_xO_2$ type or Zr does not form a compound with Ti and it occurs in modified TiO_2 in the form of partly soluble, separate phase ZrO_2. The XRD patterns are obtained for higher, calculated to 4 mol% ZrO_2, contents (Fig. 5). A weak diffraction peak at 2θ=30.3° is exhibited, indicating ZrO_2 phase in the sample (JCPDS-01-079-1771). It is more noticeable for 7.8 mol% ZrO_2 where new peaks of ZrO_2 phase at 2θ=50.4° and 2θ=51.1° are apparent. It is also observed that the increase of ZrO_2 contents resulted in the 100% peak shift to smaller angles toward higher lattice plane distances This shift informs about replacing Ti atoms with smaller ionic radius with Zr atoms with higher ionic radius, thus solid solution can be formed next to separate phase ZrO_2. The formation of either separate ZrO_2 or solid solution of Zr with Ti can restrict contact between grains and retard the anatase to rutile phase transformation.

CONCLUSIONS

Differences in the influence of studied modifiers on the anatase-rutile phase transformation resulted from different mechanisms of the reaction with TiO_2.

In our investigations we could not confirm the formation of any phase with Sb. However, Sheinkman et al. [9] showed that a certain portion of Sb^{3+} oxidised to Sb^{5+} and in reaction with TiO_2 antimony formed a new solid phase $Sb^{3+}TiSb^{5+}O_6$. Ionic radius of Ti^{4+} is 0.068 nm. Assuming that Sb^{3+} with ionic radius of 0.076 nm can increase TiO_2 unit cell parameters and Sb^{5+} with ionic radius of 0.062 nm can reduce cell parameters, this can explain why the addition of Sb_2O_3 to the TiO_2 had no effect on the anatase–rutile phase transformation.

Because of the larger ionic radii of Ce^{4+} (0.101 nm) and Zr^{4+} (0.080 nm), in comparison to Ti^{4+}, these modifiers can remain on the TiO_2 surface and form separate phases, CeO_2 or ZrO_2, which may retard the anatase-rutile phase transformation. In our investigations the inhibiting influence of CeO_2 and ZrO_2 on the anatase-rutile phase transformation is confirmed.

Zn^{2+} with ionic radius of 0.074 nm probably substitutes Ti^{4+} in the titanium dioxide crystal lattice. Moreover, ionic radius of B^{3+} is 0.020 nm, therefore, small amounts of this modifier may locate in the interstitial sites of the TiO_2 crystal lattice. Hence, these two modifiers should facilitate the rutile formation, which is consistent with the presented results.

ACKNOWLEDGMENTS

This work was funded from financial support on science for 2008–2011.

REFERENCES

1. R. J. Tayade, P. K. Suriola, R. G. Kulkarni and R. V. Jasra, *Sci. Technol. Adv. Mater.* **8**, 455 (2007).
2. W. Dąbrowski, A. Tymejczyk and A. Lubkowska, *Properties and application of titanium dioxide pigments*, Chemical Plant "Police" S.A. (Police, 2006).
3. A. V. Tolchev, V. Yu. Pervushin and D. G. Kleshchev, *Russ. J. Appl. Chem.* **75**, 696 (2002).
4. H. Ratajska, *J. Therm. Anal. Calorim.* **38**, 2109 (1992).
5. K. N. P. Kumar, *Scr. Metall. Mater.* **32**, 873 (1995).
6. U. Gesenhues and T. Rentschler, *Solid State Chem.* **143**, 210 (1999).
7. H. Zhang and J. F. Banfield, *J. Mater. Res.* **15**, 437 (2000).
8. Z. Lendzion-Bieluń and W. Arabczyk,. *Appl. Catal., A* **207**, 37 (2001).
9. A. I. Sheinkman, L. M. Goldshtein, V. N. Turlakov and G. V. Kleshchev, *Zh. Prikl. Khim.* **45**, 940 (1972).

Mater. Res. Soc. Symp. Proc. Vol. 1373 © 2012 Materials Research Society
DOI: 10.1557/opl.2012.298

Biocorrosion of AISI 4340 Steel

Sabrina M. Rovetta[1,2], Antonio J. Abdalla [2], Sonia Khouri[1], Choyu Otani[3] and Walter Miyakawa[2]
[1]Vale do Paraíba University, Av. Shishima Hifumi,2911, São José dos Campos, SP, Brazil.
[2]Photonics Division, Instituto de Estudos Avançados, Trevo Cel. Av. José A. A. Amarante, 1, São José dos Campos, SP, Brazil.
[3]Physics Department, Instituto Tecnológico de Aeronáutica, Praça Mar. Eduardo Gomes, 50, São José dos Campos, SP, Brazil.

ABSTRACT

The objective of the present work is to evaluate the *Penicillium candidum* filamentous fungi biocorrosion effects on AISI 4340 steel. Small AISI 4340 steel blocks are exposed to a biocorrosion process inside glass tubes containing culture media (Sabouraud Dextrose *HIMEDIA* broth) inoculated with *Penicillium candidum* spores for 14 days, at 25°C constant temperature. The surface microstructures are evaluated by scanning electron microscopy, atomic force microscopy, and the chemical composition by energy dispersive X-ray spectroscopy. Comparison of micrographies before and after biocorrosion shows that surface structures present morphological alterations, suggesting corrosion wear. Grain contours can no longer be visualized and oxygen content on the steel surface increases to 32% after biocorrosion. Besides, topographic parameters like root mean square roughness (Rms), arithmetic mean roughness (Ra) and mean roughness (Rz) increase 57%, 132%, and 71%, respectively, from their initial values. It is concluded that AISI 4340 steel is reasonably susceptible to corrosion.

INTRODUCTION

Metabolites produced throughout microorganisms life cycles can be highly aggressive, leading to materials biocorrosion. This deterioration process can act in several metals like aluminum and its alloys, iron, steel, copper, zinc and lead. In the course of time, major economic and structural damages have been generated, as corroded material restoration consumes approximately 20% of the worldwide produced steel.

The multiphase AISI 4340 steel is an advanced and special metal widely used in aeronautics and aerospace industries. Although high technology has been employed in its development, surface properties like hardness, corrosion resistance, friction wear, and fatigue life can be still improved by means of thermochemical treatments. Laser carburizing [1] and plasma nitrocarburizing [2,3,4] surface treatments have been effective to this aim. In fact, increases in corrosion resistance and high surface hardness have been already reported [2,4]. However, microbial induced corrosion investigations in the multiphase AISI 4340 steel are not frequently found in the literature. In a biocorrosion process, metal deterioration occurs as a consequence of an active participation of the biofouling [5], which is formed by biofilms, microbial metabolites, and corrosion products. Actually, a consortium of microorganisms is generally involved in the overall process. However, the evaluation of corrosive effects owing to a single microorganism is equally important, and from the scientific point of view, it is one of the first steps towards a more comprehensive knowledge not only of the biocorrosion mechanisms but mainly of the surface treatments that should enhance microbial induced corrosion resistance. The microorganism *Penicillium candidum* used in this work is a saprophytic filamentous fungi, typically used in dairies to make Brie and Camembert cheeses. In spite of its low pathogenicity, the corrosion potential of this fungi is comparable to *Aspergillus* genus [6,7,8]. Hence, in this work, biocorrosion effects induced by the *Penicillium candidum* filamentous fungi on the microstructure of the AISI 4340 steel are evaluated.

EXPERIMENT

Sample preparation

From a plate of AISI 4340 steel, five small blocks (5 mm x 5 mm x 3 mm) are cut. These blocks are ground on 80, 100, 240, 400, 600 and 1200 grit sandpaper, polished with diamond pastes (9 μm, 6 μm and 1 μm), and finally acid-etched with 5% Nital solution for 10 seconds. Subsequently, the surfaces of the small blocks are evaluated by SEM-scanning electron microscopy *(LEO 1460Vp, Carl Zeiss, USA)* and by AFM-atomic force microscopy *(SPM 9500J3, Shimadzu, Japan),* in contact mode, using conventional SiNi probes. Surface chemical microanalyses are also performed by EDS-energy dispersive spectroscopy *(X-MAX EDS, Oxford Instruments, UK).*

Biocorrosion

Revitalization of lyophilized *Penicillium candidum (PC-TT033, CHR HASEN, Denmark)* is performed in sterile distilled and deionized water. A pure culture (Figure 1a) is then grown in Sabouraud Dextrose agar *(Himedia, India)*, and from this pure culture, a fungal spore suspension (1.50×10^5 spores/ml) is prepared with sterile distilled and deionized water.

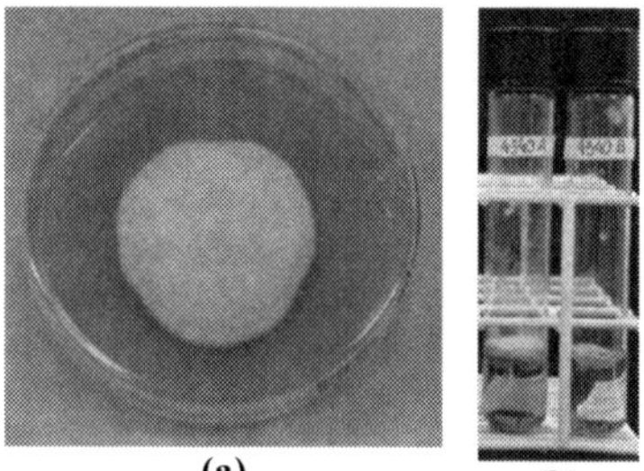

(a) **(b)**

Figure 1. (a) 15-day *Penicillium candidum* pure culture and **(b)** glass tubes containing Sabouraud Dextrose broth and the 4340 steel block, after 14 days of incubation time.

A set of glass tubes with 3 ml of Sabouraud Dextrose broth *(Himedia, India)* is inoculated with 100 μl of the *Penicillium candidum* spore suspension. In each tube, a single piece of metal block is immersed. These tubes are then kept at a constant temperature of 25°C for 14 days. After this incubation time, the *Penicillium candidum* has grown and could be visualized inside the tubes by naked-eye inspection (Figure 1b). Samples are then evaluated again using the same techniques (SEM, AFM and EDS).

RESULTS AND DISCUSSION

Figure 2a illustrates a characteristic SEM micrograph of the 4340 steel surface before biocorrosion. The observed microstructures, typical of hot rolled AISI 4340 steel, are a rather complex mixture of pro-eutectoid ferrite, upper bainite, pearlite and cementite. The cementite plates (Fe_3C, iron carbide) exhibited irregular shapes over pearlite grains and a matrix composed of bainite (darker regions) and some pro-eutectoid ferrite at grain boundaries. After biocorrosion (Figure 2b), the presence of gross corrosion products (dashed circle) and corrosion pits (white arrows) is evident. EDS-microanalyses on the corrosion products revealed high oxygen contents (~32% in weight). As oxygen had not been detected in the metal before biocorrosion, oxide formation is clearly associated with the formation of the corrosion products.

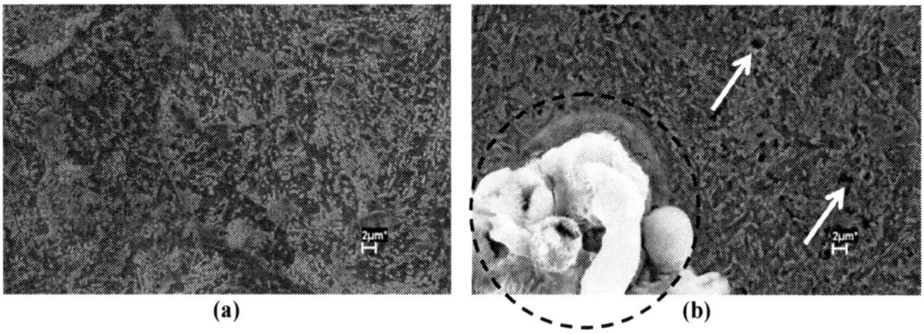

(a) **(b)**

Figure 2. SEM micrographs of 4340 steel blocks **(a)** before and **(b)** after biocorrosion. Dashed circle delimits gross corrosion structures and the white arrows show corrosion pits.

Figure 3a exhibits characteristic AFM images of the 4340 steel blocks before biocorrosion. Cementite plates over the matrix (darker regions) and pro-eutectoid ferrite plates (dashed circles) can be visualized. In Figure 3b, after biocorrosion, the preferential regions of localized pitting corrosion can clearly be observed at matrix-cementite interfaces. The height scale on the right side of each micrograph gives the vertical distance between the peak of the highest cementite plate and the lower valley of the matrix, in addition to the height distribution histogram. It can be noted that the peak-valley distance after biocorrosion is approximately twice the distance before biocorrosion. This observation indicates that, apart from severe pitting corrosion, the steel matrix surface has undergone some generalized attack.

Topographic parameters Rms (root mean square roughness), Ra (arithmetic mean roughness), and Rz (mean roughness) are also measured using the AFM off-line software facility (*Software version 2.11, Shimadzu SPM-9500 series, Shimadzu Corp. Inc., Japan*). The results showed that Rms, Ra, Rz have increased 57%, 132%, and 71%, respectively, of their initial values, reinforcing the topographical changes experienced by the 4340 steel blocks after biocorrosion.

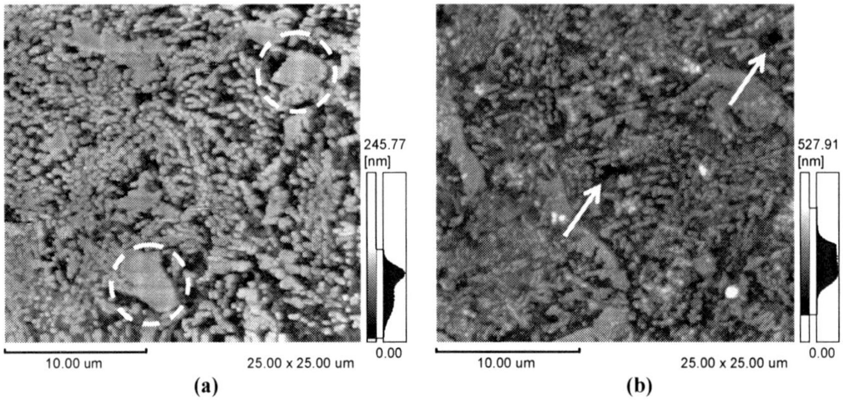

Figure 3. Contact mode AFM micrographs of 4340 steel blocks **(a)** before and **(b)** after biocorrosion.

CONCLUSIONS

The *Penicillium candidum* filamentous fungi induced morphological and topographical changes in the multiphase AISI 4340 steel. Quantitative and qualitative data obtained by SEM, AFM and EDS allowed concluding that this metal is susceptible to biocorrosion and some surface treatment is necessary to increase its biocorrosion resistance.

ACKNOWLEDGMENTS

The authors wish to thank CNPq - Conselho Nacional de Desenvolvimento Científico e Tecnológico (National Counsel of Technological and Scientific Development) for the PIBIC/CNPq-IEAv scholarship, Capes - Coordenação de Aperfeiçoamento de Pessoal de Nível Superior for the project financial support (Capes Pró-Defesa 014/08), ITA - Instituto Tecnológico de Aeronáutica, for the AFM facility, and NUFABI/UNIVAP – Núcleo de Estudos Farmacêuticos e Biomédicos da Universidade Vale do Paraíba, for the microbiology infrastructure.

REFERENCES

1. G. Vasconcelos,G. N. P. Costa, C. B. Mello, E. F. Antunes. *Mat. Sci. Forum* **660-661**, 249-251 (2010).
2. A. J. Abdalla, I. R. Carrer, M. J. R. Barboza, C. Moura Neto: Proc. of 19th CBECiMaT - Brazilian Congress on Engineering and Materials Science, ref. 317-045, Campos de Jordão, SP, 2011. Available: http://www.cbecimat.com.br/detalhes.asp?Id=886 [August 26, 2011].
3. A. Ranieri, V. H. Baggio-Scheid, P. A. Suzuki, A. J. Abdalla. Proc. of COBEM 2009 - International Congress on Mechanical Engineering **01**, 01-06, 2009.
4. A. J. Abdalla, V. H. Baggio-Scheid: *Corros. Prot. Mater.* **25**(3), 92-96 (2006).

5. H. A.Videla, *Manual of biocorrosion*. 1rst ed., CRC Press, Inc., 1996.
6. H. A.Videla, *Biotecnologia- Corrosão microbiológica*. 1rst ed., Edgard Blücher Ltda, São Paulo, 1986.
7. A. M. A. Silva, T. M. Santiago, C. R. Alves, M. I. F. Guedes, J. A. K. Freire, R. H. S. F. Vieira, R. C. B. da Silva. *Anti-corrosion Meth. Mat.* **54**, 289-293 (2007).
8. V. F. Smirnov, D. V. Belov, T. N. Sokolova, O. V. Kuzina,V. R. Kartashov. *Appl. Biochem Microbiol.* **44**, 92-196 (2008).

Mater. Res. Soc. Symp. Proc. Vol. 1373 © 2012 Materials Research Society
DOI: 10.1557/opl.2012.299

Study of the Macrosegregation of Silicon in Steels for Electrical Applications

V. A. Aranda Villada[1], J. A. García Hinojosa[1], H. Cruz Mejía[1], A. A. Balandra Aranzueta[1]., M. G. González F[1]., Y. Houbaert[2]

[1]Departamento de Ingeniería Metalúrgica, Facultad de Química, Edif. D, UNAM, Ciudad Universitaria, Coyoacán, México D. F. CP 04510;
[2]Department of Metallurgy and Materials Science, Ghent University, B-9052 Zwijnaarde, Belgium

ABSTRACT

Macrosegregation is harmful to the material properties (strength, corrosion resistance, etc.) and it must be avoided or eliminated from castings. In this work, macrosegregation behavior is evaluated in terms of silicon concentration in cast ingots of steels with 0.5, 1.5 and 3 wt% Si. The dimensions of the castings are 25x30x12.5 cm. The Si segregation profiles are evaluated in three zones: upper, middle and bottom of each ingot. The characterization is carried out by energy dispersive spectrometry in at least 10 positions of each zone of the ingot. Silicon analysis is evaluated from the surface to the centre of each ingot. The experimental results are compared with predictions using an Experimental Segregation Index (ESI) and the Scheil Segregation Index (SSI). The experimental results show a strong deviation from the profiles predicted by the Scheil model. The evaluation of the degree of segregation by "ESI" indicates little variation of silicon from the surface to the center of the ingot showing a uniform distribution across the ingot analyzed.

INTRODUCTION

Macrosegregation is a phenomenon characterized by variations of concentration in different zones of ingots and castings, particularly from their surface to the center. Macrosegregation patterns are affected by conditions prior to solidification, where gravity macrosegregation [1] and the Ludwig-Soret [2] effect are important, and during solidification. In this case, macrosegregation patterns are classified as normal [3], inverse [4,5], freckles [6], banded [7] and ingot macrosegregation [8]. In general, segregation occurs as a consequence of periodic solute enrichment or depletion, caused by the advance of the solidification front under non-stationary state conditions. These events define the magnitude of solute segregation reached at the end of solidification. In addition, fluid flow in the liquid phase and the value of the equilibrium distribution coefficient K_o of the alloy have a significant role in the evolution of macrosegregation. The chemical composition of a specific region in the ingot can be determined qualitatively or semi quantitatively by Energy Dispersive Spectroscopy (EDS) [9]. Anisotropic properties can be observed after hot rolling as result of macrosegregation in Fe-Si steels. In order to reduce or eliminate these problems, an annealing treatment is suggested before the hot rolling process.

In this paper, the composition profiles of three different steels Fe-Si (Fe-0.5%wt, Fe-1.5%wt and Fe-3%wt Si) are evaluated experimentally. The experimental results are compared with predictions performed using the Scheil–Gulliver segregation model [10].

EXPERIMENT

Iron ingots are melted in an electric induction furnace under an inert argon gas atmosphere. The chemical composition of the melt to the desired Si content is adjusted with ferrosilicon (FeSi75). The Fe-Si steel alloys are poured into metallic moulds in order to obtain ingots with a total weight of 100 kg including the feeding system, useful casting weight is 65 kg. Ingot dimensions are 25x12.5x30 cm. All alloys are solidified and cooled under similar conditions. The Si concentrations of the experimental alloys are 0.5, 1.5 and 3 wt% Si. Chemical composition of the ingots is determined by Atomic Emission Spectrometry and the results are shown in Table I.

Table I. Chemical composition of the experimental alloys (wt%)

Alloy	Chemical composition*, wt %							
	C	Si	Mn	P	S	Al	N	Cr
0.5	0.005	0.45	0.066	0.005	<0.001	0.002	0.051	0.022
1.5	0.003	1.52	0.042	0.005	0.002	0.002	0.0022	0.015
3.0	0.006	2.99	0.036	0.002	0.005	0.003	<0.001	0.014

*Fe Balance

After casting, each ingot is cut into plates of 12.5 cm width, 2.5 cm thickness and 30 cm high and each plate is cut into 6x2x2.5 cm slices. Each slice is divided into three different zones identified as upper, middle and bottom, as can be seen in Figure 1. Experimental slices are prepared by conventional metallographic techniques and etched with 1:3 fresh solution of HNO_3-H_2O in order to obtain the silicon composition profile from the surface to the center of the ingot.

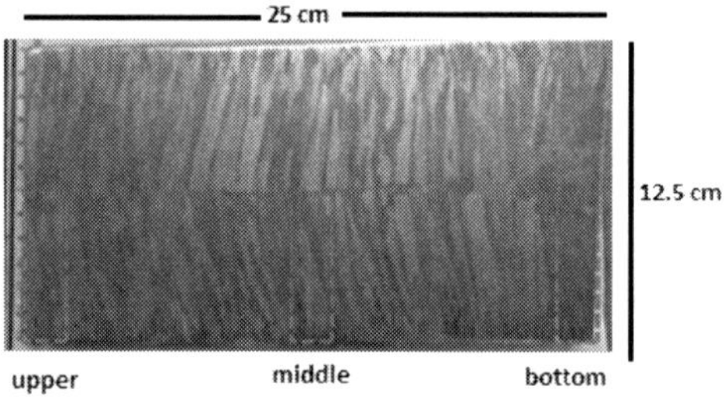

Figure 1. Zones considered in the evaluation of the silicon composition profile.

Silicon macrosegregation experimental profiles are determinate by scanning electron microscopy (SEM) coupled with an energy dispersive X-ray spectroscopy analyzer (EDS). This analysis is carried out in small specimens of 6x2x2.5 cm corresponding to the ingot positions mentioned previously. At least eleven experimental zones are considered for this analysis; each zone had a separation distance of 5 mm, approximately. The experimental analysis is made from the edge through the center of the sample as can be seen in Figure 2; the approximate analysis area is 598 x 598 μm.

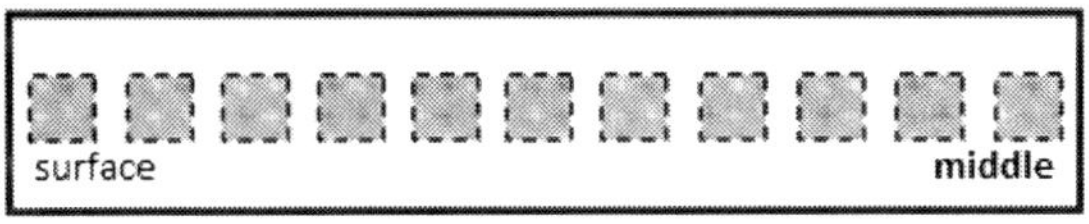

Figure 2. Positions corresponding to each one of the zones analyzed.

Sheil-Gullivert model is applied in order to predict the macrosegregation in the alloys under study. This model assumes complete mixing in the liquid phase and a no mixing in the solid phase. Because of the low alloy content and the columnar microstructures observed in two of three alloys under study, the alloys are considered as binary systems for modeling. According to the Scheil model, the solute concentration in the solid, C_s, in a certain position in the ingot where the solid fraction is f_s, is given by:

[illegible] (1)

In Ec. 1, C_o is the Si concentration in the original alloy and, K_o is the solute distribution coefficient. Finally, experimental and Scheil segregation indexes (ESI, and SSI respectively) are calculated according to:

[illegible] (2)

Where C_{max} is the maximum Si concentration when f_s=0.9 and, C_{min} is the minimum concentration of silicon. Both concentrations can be related to the experimental results or linked to the predictions obtained by Scheil model.

RESULTS AND DISCUSSION

Figures 3, 4 and 5 show the average experimental profiles for the 0.5, 1.5 and 3% Si alloys, respectively. A comparison is made with the predictions obtained using the Scheil equation. As can be seen, all three alloys exhibit irregular Si distribution profiles. However, the compositional variations observed along the ingot are of little significance with respect to the nominal composition of the alloys. This effect can be attributed to the form adopted by the interface during solidification of the alloy and the diffusive-convective conditions prevailing in the liquid–solid front. As shown by comparing Figures 3-5, this effect is more pronounced, in the alloys with 1.5 and 3%wt Si. In contrast, the predictions made by the Scheil model show a gradual increase of Si from the surface to center of the ingot. These calculations are performed assuming $K_o < 1$. The values considered for the solute distribution coefficient are extracted from the binary Fe-Si phase diagram. These are 0.43 for Fe-0.5%wt Si alloy, 0.48 for Fe-1.5%wt Si alloy and 0.62 in the case of Fe-3%wt Si alloy.

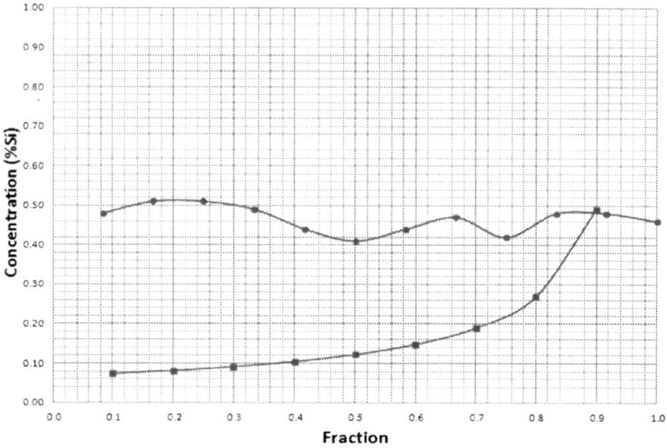

Figure 3. Fe-0.5 wt% Si alloy macrosegregation silicon profile, experimental (•) and Scheil model profile (■)

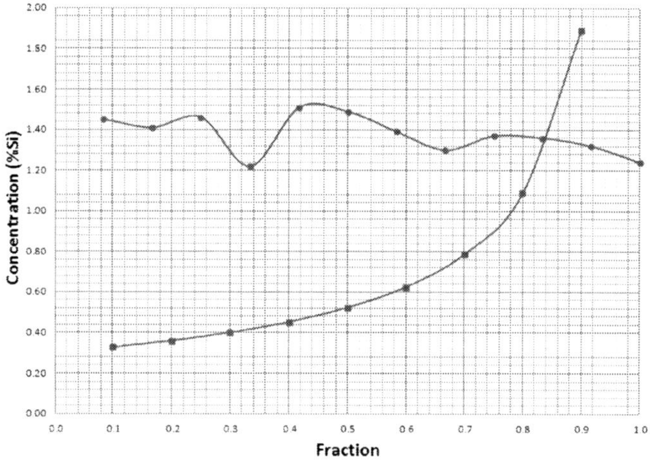

Figure 4. Fe-1.5 wt% Si alloy macrosegregation silicon profile, experimental (•) and Scheil model profile (■)

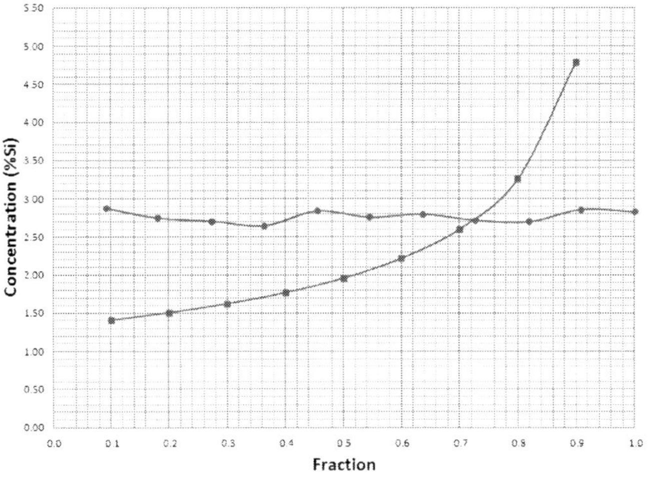

Figure 5. Fe-3.0 wt% Si alloy macrosegregation silicon profile, experimental (•) and Scheil model profile (■).

Table II lists the segregation index values obtained from the experimental results and those calculated from the Scheil model predictions corresponding to each zone of the three alloys analyzed in this study.

Table II. Experimental segregation index (ESI) and Sheil segregation index (SSI) associated of each zone analyzed.

Alloy	Ingot zone	ESI	SSI
Fe-0.5%Si	Upper	1.13	3.7
	Middle	1.38	
	Bottom	1.24	
Fe-1.5 % Si	Upper	1.10	3.3
	Middle	1.06	
	Bottom	1.07	
Fe-3%Si	Upper	1.09	2.4
	Middle	1.08	
	Bottom	1.08	

Comparatively, according with these results, the largest compositional variation is present in the Fe-0.5% wt Si alloy. Alloys with larger Si concentration resulted in smaller values of segregation index. These results suggest that the homogenization treatment required for ingots of

Fe-0.5%wt Si alloy must be carried out during longer times and at higher temperatures in comparison with the other two alloys under study.

CONCLUSIONS

The segregation index in solidified Fe-Si alloy ingots decreases as the Si content increases. Fe 0.5% Si alloy exhibits the highest segregation index while alloys with higher Si concentrations exhibit only a small variation of this element from the surface to the center of the ingot. The experimental segregation patterns show a strong deviation from the predictions of the Scheil model; nevertheless they follow the same general trend predicted by the model. The low segregation index observed in the 1.5 and 3 % Fe-Si alloy ingots is attributed to the strongly columnar solidification pattern observed in these alloys.
The magnitude of the segregation index is important in order to set the annealing heat treatment conditions before the rolling process.

ACKNOWLEDGMENTS

This work is supported by ACERLECT FONCICYT-CONACYT No. 93325 in collaboration with UANL Universidad Autónoma de Nuevo León Dr. Rafael Colás y Dra. Martha Guerrero. Also thank to IQM E. Candido Atlatenco T. for his help of the sample analysis.

REFERENCES

1. I. L. Ferreira, J. E. Spinellia and A. Garcia, J. Alloys Compd., **475**, 396 (2009).
2. B. Prabhakar, M. S Thesis, University of British Columbia, 1973
3. Y. L Liu, and S. B Kang, Mater. Sci. Technol., **13**, 331 (1997)
4. H. Kato, J.R. Cahoon, Metal. Mater Trans. A, **16**, 579 (1985)
5. M.C. Flemings and G.E. Nereo, Trans TMS-AIME, **239**, 1449 (1967)
6. W. Yang, W. Chen, K. Minn, S. Mannan and J. de Barbadillo, Metal. Mater Trans. A, **32**, 397 (2001)
7. W. Kurz, and R. Trivedi, Metal and Mater Trans A, **27**, 625, (1996)
8. J.P Gu and C. Beckermann Metal Mater Trans A, **30**, 1357, (1999)
9. González M. G., Noguez A. M. E., Principios de Microscopía Electrónica de Barrido y Microanálisis por Rayos X,(UNAM, Facultad de Química,2006), 120-134.
10. T.P. Battle and R.D. Pehlke, Metal and Mater Trans B, **21**, 357 (1990)

Mater. Res. Soc. Symp. Proc. Vol. 1373 © 2012 Materials Research Society
DOI: 10.1557/opl.2012.300

Effect of Microalloying Elements (B, Nb, V and Ti) on the Strain Hardening Behavior of High-Manganese TWIP Steels.

F. Reyes-Calderón[1], I. Mejía[1] and J.M. Cabrera[2,3]

[1]Instituto de Investigaciones Metalúrgicas, Universidad Michoacana de San Nicolás de Hidalgo, Edificio "U-5", Ciudad Universitaria. 58060-Morelia, Michoacán. México.
fpacor@yahoo.com.mx, imejia@umich.mx.
[2]Departament de Ciència del Materials i Enginyeria Metal·lúrgica, ETSEIB, Universitat Politècnica de Catalunya, Av. Diagonal 647. 08028-Barcelona, Spain.
jose.maria.cabrera@upc.edu.
[3]Fundació CTM Centre Tecnològic, Av. de les Bases de Manresa 1. 08242-Manresa (Barcelona), Spain.

ABSTRACT

The present research work analyses the influence of microalloying elements (B, Nb, V and Ti) on the tensile strength and the strain hardening behavior of a high-manganese TWIP steel. The analysis was carried out by means of true stress-true strain curves derived from uniaxial tension tests. The work hardening exponent was determined by using the Hollomon and differential Crussard-Jaoul models. Metallographic characterization was carried out to determine the metallurgical changes associated with *n* values. The results indicate that the Hollomon analysis results in strain hardening exponent values up to 0.46. On the other hand, the differential Crussard-Jaoul analysis establishes a clear distinction of *n* value for two stages of plastic deformation which are determined by a sharp slope change in the plot of $ln(d\sigma/d\varepsilon)$-$ln\varepsilon$.

INTRODUCTION

TWIP steels have been widely considered for many industrial applications, particularly in the automotive sector. These steels can provide both high strength and large ductility because twins can produce plastic deformation and, at the same time, twin boundaries act as an obstacle for dislocation movement [1]. The occurrence of twinning depends strongly on the stacking fault energy (SFE), which in turn depends on chemical composition and temperature [2]. The twinning effect produces a high value of the strain hardening exponent *n*. Xiong et al. [3], investigated the dynamic tensile properties of TWIP steels and determined that *n* increases with strain rate due to formation of twins during deformation, which then become an obstacle for dislocation slip. The strain hardening evolves with deformation such that: (i) at the beginning of plastic yielding deformation by dislocation slip is the dominant deformation mode and *n* decreases; (ii) once the stress required to initiate twinning is reached, primary twins with similar orientations proliferate thereby increasing the *n* value at a constant rate; (iii) at even higher stresses, secondary twins are formed and intersect primary twins and further increases the *n* exponent. When secondary twinning occurs while primary twinning is still active, the two mechanisms overlap and this causes a gradual decrease in *n* value [4, 5]; (iv) further straining of previously formed twins hinder the production of new twins thereby decreasing *n*. It must be emphasized that the relative extent of each of these strain hardening regions depends on the SFE, grain size and crystallographic orientation [4, 6-9]. The main goal of the present research work is to determine the influence of the microalloying elements (B, Nb, V and Ti) on the strain hardening behavior of Fe-Mn-Al-Si-C TWIP steels.

Analytical methods

The Hollomon analysis. The tensile deformation behavior of polycrystalline metals in the uniform deformation regime is explained by suitable constitutive stress-strain relationships such as the classical Hollomon [10] equation:

$$\sigma = k_H \cdot \varepsilon^{n_H} \quad (1)$$

σ is the true stress, ε is the true strain, k_H and n_H are experimental constants obtained from the slope and intercept of a $ln\sigma$-$ln\varepsilon$ plot, respectively.

The differential Crussard-Jaoul (C-J) analysis. The differential C-J analysis of the σ–ε curves of metals and alloys is quite useful to determine significant variations of the strain hardening exponent. The Ludwick power relationship [11] is assumed:

$$\sigma = \sigma_0 + k \cdot \varepsilon^n \quad (2)$$

σ_o is the yield stress, k is the strength coefficient and n the strain hardening exponent. The logarithmic form of the equation 2, after differentiation with respect to ε, formally establishes the analysis method.

$$\text{In}\frac{d\sigma}{d\varepsilon} = \text{In}\left(k \cdot n_{C\text{-}J}\right) + \left(n_{C\text{-}J} - 1\right) \cdot \text{In}\varepsilon \quad (3)$$

In a $ln(d\sigma/d\varepsilon)$-$ln\varepsilon$ plot, the slope of the line gives the ($n_{C\text{-}J}$ -1) value, while the intercept at $ln=0$ provides the $ln(kn_{C\text{-}J})$ value. The derivation of $n_{C\text{-}J}$ and k is then straightforward. Also σ_o can be estimated by fitting Ludwick's equation 2 to the experimental data.

EXPERIMENTAL DETAILS

Five TWIP steel ingots, with a base chemical composition of Fe-20Mn-1.5Al-1.5Si-0.4C (wt%) were melted in an induction furnace. One of the steel ingots, designated (TWIP-NM) was not microalloyed while the other four were single microalloyed with Boron (TWIP-B), Niobium (TWIP-Nb), Vanadium (TWIP-V), and Titanium (TWIP-Ti). Samples from the ingots were homogenized during 5 hours at 1200ºC, and then hot rolled up to a 60% reduction of area. After that, the samples were quenched to room temperature in water.

Reduced ASTM-E8 [12] tensile test specimens were machined from the hot rolled plates and standard tensile tests were carried out up to strain-to-failure at a strain rate of $10^{-1}s^{-1}$ in an INSTRON-5585H universal testing machine. The load-elongation data in the uniform deformation range was processed to produce true σ-ε data using an adjacent averaging smoothing routine. Finally, the n values were determined by linear regression using a least square method in the logarithmic plots in both models. Metallographic preparation was done using standard techniques and etching was carried out with 10% Nital solution. The metallographic results were obtained using light optical microscopy (LOM). The X-ray diffraction patterns (XRD) were recorded in a diffractometer using CuKα radiation at a scanning rate of 0.02° s^{-1} between 30 and 120 °2θ.

DISCUSSION

Figure 1(a) shows the typical stress-strain curve obtained for the TWIP-NM steel. Similar curves were obtained for the other four TWIP steels. Figure 1(b) shows the results of the Hollomon analysis where a good general average fitting to the experimental data is observed. The differential C-J analysis is more sensitive to slope changes in the true σ-ε curve and results in a much more better fit to the experimental data, as shown in Figure 1(c), here two different strain hardening stages can be noticed.

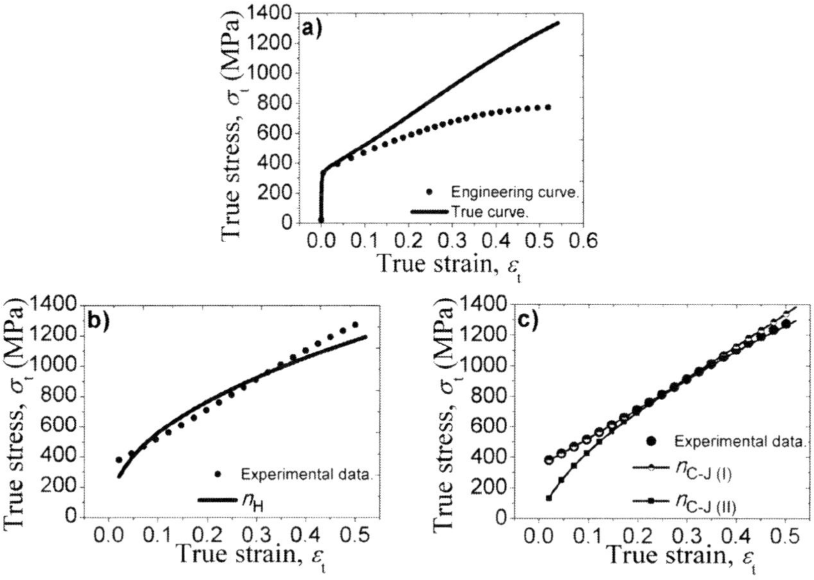

Figure 1. Stress-Strain behavior of TWIP-NM steel: a) Stress-strain curves; b) Hollomon analysis; c) differential Crussard-Jaoul analysis.

The strain hardening exponents and constants calculated using both the Hollomon and differential C-J analysis methods are listed in Table I. As can be seen, n_H ranges from 0.38 in the TWIP-B steel up to 0.46 in the TWIP-NM and TWIP-Nb steels. The $n_{C\text{-}J}$ values ranges from 1.03 to 1.10 in the first stage and from 0.61 to 0.92 for the second deformation stage.

Table I. Work hardening exponents determined by Hollomon and differential Crussard-Jaoul models for microalloyed and no microalloyed TWIP steels.

TWIP steel	True uniform strain	Hollomon Analysis. n_H	k_H	Differential C-J analysis $n_{C\text{-}J}$	$k_{C\text{-}J}$	σ_0
NM	0.0205-0.2379	0.46	1609.6	1.08	2093.1	348.7
	0.2379-0.5210			0.61	2010.0	-56.8
B	0.0171-0.2895	0.38	1373.6	1.08	1980.5	359.2
	0.2895-0.3865			0.92	1905.3	268.6
Nb	0.0252-0.2037	0.46	1621.5	1.05	2048.9	363.1
	0.2037-0.4783			0.72	1857.3	148.5
V	0.0235-0.2725	0.45	1593.6	1.10	2046.9	368.6
	0.2725-0.5069			0.71	1964.7	68.3
Ti	0.0196-0.1578	0.43	1555.6	1.03	2054.0	346.4
	0.1578-0.4610			0.74	1792.6	186.2

Figure 2 shows optical micrographs of TWIP-NM steel before and after the tensile test. Similar microstructures were observed for the other microalloyed TWIP steels. It is evident the presence of annealing and deformation twins in some grains before the tensile test. The average grain size in the annealed condition was 104 μm, Figure 2(a). The annealing twins have a relatively larger thickness compared with those produced after the tensile test, Figure 2(b), in which the elongated grains have a high thin twin density. The twin lamellae distance measured in a parallel study using SEM-EBSD is between 0.4-0.6 μm. It is possible to observe some mechanical and annealing twins, which act as grain boundaries blocking the twin propagation and define different twinned zones in elongated grains, region A in Figure 2(b). In addition, some grains show different twinning system activated, region B in Figure 2(b), as previously reported Soulami et al. [2]. The contrast in the micrograph of Figure 2(b), is due to the texture evolution produced during cold deformation. There are few white zones corresponding to twin-free austenite, region C in Figure 2(b).

XRD patterns before and after tensile deformation for the five TWIP steels investigated are presented in Figure 3. As can be seen prior to deformation, the microstructure consists mainly of strongly textured FCC austenite with a small amount of ε-martensite. The martensitic transformation is common in steels that present TRIP or TRIP-TWIP effect because the austenite phase is not fully stabilized during the solution heat treatment [13]. After tensile deformation to fracture there is a significant overall decrease in intensity, the peaks associated with the martensite phase disappear and the diffraction peaks become sharper, particularly in the TWIP-B steel, Figure 3(b). This later effect can be attributed to the crystal re-orientations caused by the formation of deformation twins [14].

Many researchers [7, 8, 15-20] have demonstrated that formation of deformation twins leads to a strong increase in strain hardening. The results of the differential C-J analyses (Figure 1 and Table I) of the present experimental data and the microstructural evidence in Figures 2 and 3, reveals that this is also the case for the experimental steels investigated in this work. The increase in strain hardening rate is usually attributed to a reduction of the dislocation slip mean free path which results from the twin boundaries acting as obstacles to dislocation glide. This effect causes the observed increase in the strain hardening exponent and indicates that twinning strengthening plays an important role in strain hardening in the present materials.

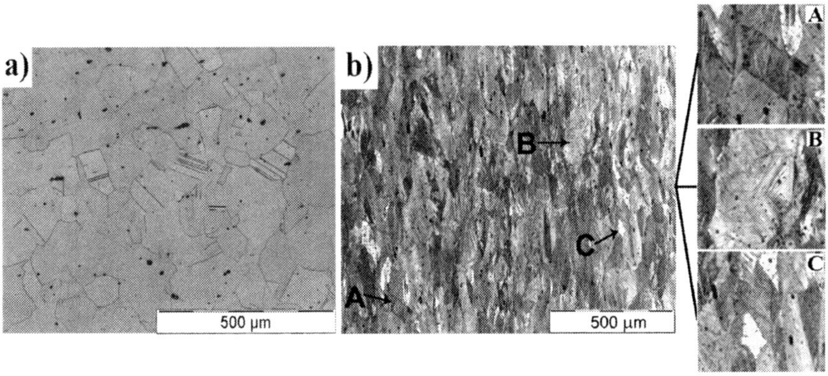

Figure 2. Microstructure of TWIP-NM steel; a) Solution treated condition, b) cold deformed condition.

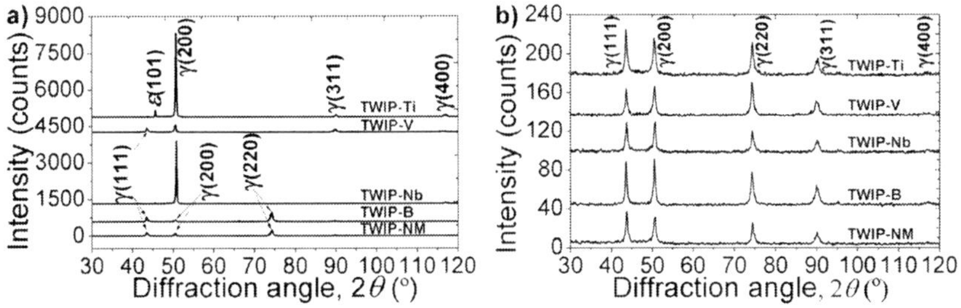

Figure 3. XRD patterns of TWIP steels; a) solution heat treated condition, b) after tensile deformation to fracture.

CONCLUSIONS

The results of the present work show that the Hollomon strain hardening exponent is nearly constant with values from 0.43 to 0.46 and independent of the type of microalloying addition (Nb, V and Ti) for the TWIP steels investigated. A smaller value of n_H (0.38) was observed for the TWIP-B steel. The differential C-J analysis fits better the experimental data and shows clearly two stages of strain hardening indicating a change in the plastic deformation mechanism. Precipitate particles containing microalloying elements (B, Nb, V and Ti) do not seem to have a significant effect on strain hardening of the investigated steels. On the basis of the present results, the TWIP effect is responsible for the observed changes in strain hardening.

ACKNOWLEDGMENTS

The present project has been financially and technically supported by the Universitat Politècnica de Catalunya, the Centre Tecnològic de Manresa, (Spain), and the Instituto de Investigaciones Metalúrgicas-UMSNH (México). Reyes-Calderon's studies are sponsored by CONACYT (México), (No. CVU: 297577). The authors want to give a special acknowledgement to Eng. Carles Abad Olesa and Mr. Marc Pla Pardo from CTM, Manresa, for their help in the tensile tests realization.

REFERENCES

1. O. Bouaziz, and N. Guelton, *Mater. Sci. Eng. A.* **319-321**, 246-249 (2001).
2. A. Soulami, K. S. Choi, Y. F. Shen, W. Liu, X. Sun, and M. A. Khaleela, *Mat. Sci. Eng. A.* **528**, 1402-1408 (2011).
3. R. Xiong, R. Fu, Y. Su, Q. Li, X. Wei, and L. Li, *ISIJ Int.* **16**, 81-86 (2009).
4. G. Dini, A. Najafizadeh, R. Ueji, and S. M. Monir-Vaghefi, *Mat. & Des.* **31**, 3395-3402 (2010).
5. J. A. Jiménez, and G. Frommeyer, *Mat. Charact.* **61**, 221-226 (2010).
6. R. Ueji, N. Tsuchida, D. Terada, N. Tsuji, Y. Tanaka, A. Takemura, and K. Kunishige, *Scripta Mat.* **59**, 963-966 (2008).
7. S. Asgari, E. El-Danaf, S. R. Kalidindi, and R. D. Doherty, *Metall. Mater. Trans. A.* **28**, 1781-1795 (1997).
8. E. El-Danaf, S. R. Kalidindi, and R. D. Doherty, *Metall. Mater. Trans. A.* **30**, 1223-1233 (1999).
9. I. Gutierrez-Urrutia, S. Zaefferer, and D. Raabe, *Mat. Sci. Eng. A.* **527**, 3552-3560 (2010).
10. J. H. Hollomon, *AIME Trans.* **12,** 1-22 (1945).
11. G. E. Dieter in *Mechanical Metallurgy*. SI metric ed. (ed. McGraw Hill, London, 1986) pp. 103-240.
12. ASTM Standard E8. 2003. Test Meth for Tension Testing of Metallic Materials. ASTM International, A., West Conshohocken, PA. (2003)
13. M. Koyama, T. Sawaguchi, T. Lee, C. S. Lee, and K. Tzusaki, *Mat. Sci. Eng. A.* **528**, 7310-7316 (2011).
14. D. Li, Y. Wei, C. Lid, L. Hod, D. Lid, and X. Jin, *J. Iron & Steel Res. Int.* **17**, 67-73 (2010).
15. L. Remy, *Acta Metal.* **26**, 443-451(1978).
16. J. M. Han, C. Y. Lim, and Y. G. Kim, *Acta Metall. Mater.* **39**, 2169-2175 (1991).
17. S. Allain, J. P. Chateau, and O. Bouaziz, *Steel Res. Int.* **73**, 299-302 (2002).
18. S. Allain, J. P. Chateau, and O. Bouaziz, *Mat. Sci. Eng. A.* **387-389**, 143-147 (2004).
19. S. Allain, J. P. Chateau, D. Dahmoun, and O. Bouaziz, *Mat. Sci. Eng. A.* **387–389**, 272–276 (2004).
20. S. Allain, J. P. Chateau, O. Bouaziz, S. Migot, N. Guelton, *Mat. Sci. Eng. A.* **387-389,** 158-162 (2004).

Mater. Res. Soc. Symp. Proc. Vol. 1373 © 2012 Materials Research Society
DOI: 10.1557/opl.2012.301

Mechanical properties and microstructure of low carbon ultra-high strength steels (UHSS) microalloyed with boron

I. Mejía[1], A. García de la Rosa[1], A. Bedolla-Jacuinde[1] and J.M. Cabrera[2,3]

[1]Instituto de Investigaciones Metalúrgicas, Universidad Michoacana de San Nicolás de Hidalgo, Edificio "U-5", Ciudad Universitaria. 58060-Morelia, Michoacán. MÉXICO. imejia@umich.mx
[2]Departament de Ciència del Materials i Enginyeria Metal·lúrgica, ETSEIB, Universitat Politècnica de Catalunya, Av. Diagonal 647. 08028-Barcelona, SPAIN.
[3]Fundació CTM Centre Tecnològic, Av. de les Bases de Manresa 1. 08242-Manresa (Barcelona), SPAIN. jose.maria.cabrera@upc.edu

ABSTRACT

The aim of this research work is to study the effect of boron addition on mechanical properties and microstructure of a new family of low carbon NiCrVCu advanced high strength steels (AHSS). Experimental steels are thermo-mechanically processed (TMP) (hot-rolled+quenched). Results show that the microstructure of these steels contains bainite and martensite, predominantly, which nucleate along prior austenite grain boundaries (GB). On the other hand, tensile tests reveal that the TMP steels have YS (0.2% offset) of 978 MPa, UTS of 1140 MPa and EL of 18%. On the basis of exhibited microstructure and mechanical properties, these experimental steels are classified as bainitic-martensitic complex phase (CP) advanced ultra-high strength steels (UHSS).

INTRODUCTION

Recent years have seen many developments in steel technology and manufacturing processes to build vehicles of reduced weight and increased safety. For this purpose Advanced High Strength Steels (AHSS) have been developed. The AHSS include newer types of steels such as dual phase (DP), transformation-induced plasticity (TRIP), complex phase (CP), B steels (BS), and martensitic steels (MART), which are primarily multi-phase steels, and contain ferrite, martensite, bainite, and/or retained austenite in quantities enough to produce outstanding mechanical properties [1-2]. Researchers have studied the B effect in steels for a long time [3-7] particularly due to its potential to increase steel hardenability. Nowadays, with the development of modern steelmaking technologies, the technique for controlling B additions is mature and B steels are extensively applied in the construction of heavy machinery, building structures, marine platforms and pipelines [8-11]. It has been well-established that B atoms segregate towards austenite GB and increase hardenability of steel by suppressing the nucleation of ferrite. There are four mainly explanations about the B effect mechanism [12]: (i) B segregation to austenite GB reduces the grain boundary energy, and so the number of preferential nucleation sites for ferrite, (ii) B reduces the self-diffusional coefficient of iron at GB and decreases the nucleation rate of ferrite, (iii) GB are a preferential nucleation sites for ferrite and when B segregates to GB, these sites will vanish, and (iv) Fine borides form along the boundaries and are coherent with the matrix; in this case, it is hard to nucleate ferrite at the boride-matrix interface. It is common for all of these explanations that equilibrium B segregation at austenite GB influences the nucleation process of ferrite but it does not affect the thermodynamic characteristics of austenite or ferrite.

Hardenability of B steels is determined by non-equilibrium B segregation on GB [12]. On the other hand, grain refinement is known as the best way of improving both strength and toughness [13]. There are several methods for grain refinement of steels such as controlled rolling, accelerated cooling and cyclic thermo-mechanical treatment [14-15]. Nowadays, there are just a few studies strictly focused on the B effect on microstructure and mechanical properties of low carbon AHSS. The present research work studies the effect of B on the microstructure obtained after thermo-mechanical processing and the resulting mechanical properties of a new family of low carbon NiCrVCu advanced ultra-high strength steels (UHSS).

EXPERIMENTAL DETAILS

The present experimental low carbon NiCrVCu advanced UHSS are melted in the Foundry Laboratory of the Metallurgical Research Institute-UMSNH using high purity raw materials in a 25 kg capacity induction furnace. The liquid steel is cast into 70x70 mm cross section ingots. Table 1 shows the chemical composition of the six experimental steels examined in this study. Before thermo-mechanical processing, the steel ingots are cut and austenitized at 1200°C for 3 h. The thermo-mechanical treatment is carried out by a reversed multipass process to reach a level of deformation of 80% in a 50 t T.J. Pigott laboratory rolling mill.

Table 1. Chemical composition of low carbon advanced UHSS microalloyed with B (wt. %).

Steel	C	Mn	Si	S	Cu	Cr	Ni	V	Al	N	B
B0	0.15	0.40	0.42	0.02	0.52	1.31	2.44	0.22	0.0026	0.0091	**0**
B1	0.12	0.40	0.40	0.02	0.51	1.31	2.38	0.22	0.0040	0.0100	**0.0014**
B2	0.11	0.41	0.43	0.01	0.46	1.33	2.26	0.24	0.0048	0.0082	**0.0033**
B3	0.11	0.40	0.35	0.02	0.51	1.30	2.37	0.22	0.0030	0.0086	**0.0082**
B4	0.10	0.40	0.33	0.01	0.49	1.30	2.30	0.22	0.0036	0.0079	**0.0126**
B5	0.10	0.41	0.32	0.02	0.50	1.30	2.42	0.22	0.0031	0.0087	**0.0214**

Plastic deformation is finished at 950°C for all steels and the plates are then water quenched to room temperature. Cylindrical tensile test specimens of 6.35 mm in diameter and 25.4 mm in gauge length are machined from the TMP steels according to ASTM E 8 standard specification. Tensile tests are carried out using an Instron tensile testing machine at room temperature and a crosshead speed of 0.01 $mm \cdot s^{-1}$. Work hardening exponent (n) is determined using Hollomon model. The analysis only considered the portion of true σ-ε curve that exhibits work hardening and n values are determined by linear regression of logarithmic plots. Characterizations of steels are undertaken by light optical (LOM) and scanning electron microscopy (SEM). Samples for LOM and SEM are prepared by the typical metallographic procedure: grinding on abrasive paper followed by polishing on nylon cloths using diamond paste as abrasive and finally etched with 2% Nital.

RESULTS AND DISCUSSION

Microstructure

Figure 1 shows the optical micrographs of the experimental TMP low carbon advanced UHSS microalloyed with B. The microstructure contains predominantly bainite and martensite, which has nucleated along prior austenite GB.

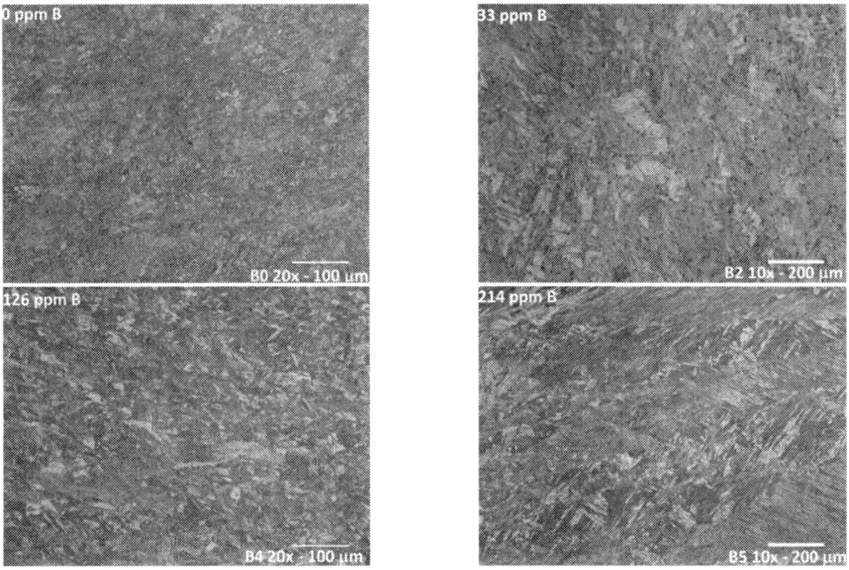

Figure 1. Optical micrographs of the TMP low carbon advanced UHSS microalloyed with B.

Figure 2 shows the SEM microstructures of the steels having considerable amount of bainite and martensite with lath morphology. As shown from Figures 1 and 2, bainite and martensite increases with increasing B content, which demonstrates that B additions retard the austenite-ferrite transformation. It is well recognized that a small amount of B dramatically enhances the hardenability of steels by preventing the nucleation of ferrite due to the B atoms segregated to the austenite GB [12]. The resultant microstructures are finer by B addition and deformation in the austenite region. In general, the LOM and SEM observations reveal a significant evolution of the microstructure with the B addition. In this case, the bainite and martensite lath morphology becomes more pronounced, larger packets of laths and longer laths can be seen in the steels with higher B content. It is worth mentioning that the martensitic lath structure obtained on continuous cooling is coarser than that obtained in TMP steels after water quenching. On the other hand, Cu addition to the present steels enhances hardenability and lowers the transformation temperature of austenite to ferrite by increasing the stability of austenite and by causing solid solution and precipitation hardening.

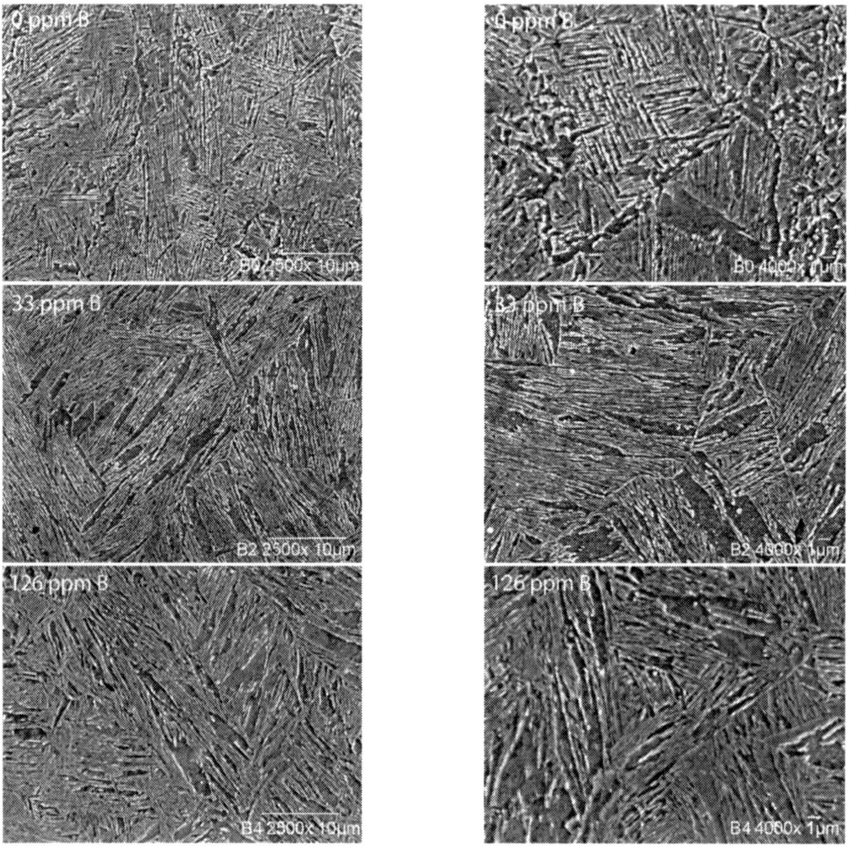

Figure 2. SEM micrographs of the TMP low carbon advanced UHSS microalloyed with B.

Mechanical properties

Figure 3 shows the mechanical properties of the TMP low carbon advanced UHSS microalloyed with B. Figure 3b shows that the UTS and YS increase as B content increases up to saturation value around 82 ppm B; on the contrary, EL to fracture tends to decrease. Results reveal that TMP steels have YS (0.2% offset) of 978 MPa, UTS of 1140 MPa and EL of 18%. On the basis of the exhibited microstructure and mechanical properties, these experimental steels can be classified as bainitic-martensitic complex phase (CP) UHSS. In this case, the higher strengths found in these steels are attributed to a combination of (i) reduced grain size and larger bainite and martensite volume fractions, (ii) fine lath martensite structure, (iii) very effective precipitation hardening, and (iv) very high dislocation densities [16-17]. In general, the major contribution of the high strength obtained in the present TMP steels is mainly due to the fine lath martensite structure, containing a high dislocation density with dispersed tiny microalloying precipitates formed on dislocations. Additionally, it is worth mentioning that the Cu addition clearly increases the YS and UTS in these steels because of its solid solution strengthening effect as well as the microstructural change caused by the enhanced hardenability.

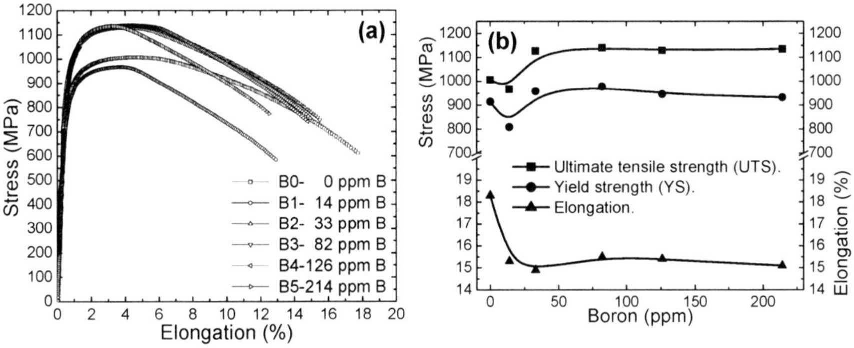

Figure 3. Mechanical properties of the TMP low carbon advanced UHSS microalloyed with B. (a) Engineering σ-ε curves. (b) UTS, YS and EL (%) as a function of B content.

Figure 4 shows the work hardening exponent behavior of the TMP low carbon advanced UHSS. In this case, Hollomon model exhibited a good fit to the experimental data. As observed from Figure 4, n_H values are low and increase (from 0.108 to 0.138) as B content increases. In general, these steels show n_H values lower than those observed for the conventional carbon steels, which is related to the different volume fractions of bainite and martensite and particularly to fine lath martensite structure, containing a high dislocation density with dispersed tiny precipitates.

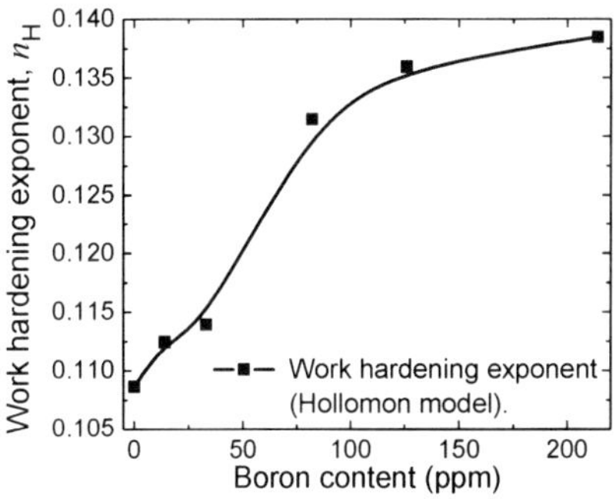

Figure 4. Work hardening exponent (Hollomon model) of the TMP low carbon advanced UHSS microalloyed with B.

CONCLUSIONS

The present TMP low carbon advanced UHSS show bainite and martensite with lath morphology microstructures. Bainite and martensite increase with increasing B content, which demonstrates that decomposition kinetics of austenite to ferrite is delayed due to the higher B content and therefore steel hardenability increases. The present TMP steels can have YS (0.2% offset) of 978 MPa, UTS of 1140 MPa and EL of 18%. These steels can be classified as bainitic-martensitic complex phase (CP) advanced UHSS. The higher strengths found in these steels are associated with a reduced grain size, higher bainite and martensite volume fractions, fine lath martensite structure, very effective precipitation hardening, and very high dislocation densities.

ACKNOWLEDGMENTS

A. García de la Rosa would like to thank CONACYT (México) for the scholarship support during this project. All the authors also acknowledge CMEM-UPC (Spain), for the support and technical assistance in this research work. Funding is obtained through project CICYT-MAT2008-06793-C02-01 (Spain) and CIC-UMSNH (México).

REFERENCES

[1] Committee on Automotive Applications, International Iron & Steel Institute, *Advanced High Strength Steel Application Guidelines* 1-9 (2006).
[2] H. -T. Jiang, D. Tang and Z. -L. Mi, *J. Iron Steel Res.* **19(8)**, 1-6 (2007).
[3] K. A. Taylor and S. S. Hansen, *Metall. Mater. Trans.* **A21**, 1697-1708 (1991).
[4] H. Tameiro, M. Murata, R. Habu and M. Nagumo, *Trans. Iron Steel Inst. Jpn.* **27**, 120-129 (1987).
[5] J. E. Morral and T. B. Cameron, in *Boron in Steel* edited by S. K. Banerji and J. E. Morral, (The Metallurgical Society of AIME, Milwaukee, USA, 1980) pp. 19-32.
[6] R. Habu, M. Miyata, S. Sekino and S. Goda, *Trans. Iron Steel Inst. Jpn.* **18**, 492-500 (1978).
[7] D. H. Werner, *"Boron and Boron Containing Steels"*, 2nd ed. (Verlag Stahl Eisen, Dusseldorf, 1995) pp. 15-20.
[8] B. M. Kapadia, *J. Heat Treat.* **5**, 41-53 (1987).
[9] C. J. Heckmann, D. Ormston, F. Grimpe, H. -G. Hillenbrand and J. -P. Jansen, *Ironmaking & Steelmaking* **32**, 337-341 (2005).
[10] H. J. Jun, J. S. Kang, D. H. Seo, K. B. Kang and C. G. Park, *Mater. Sci. Eng.* **A422**, 157-162 (2006).
[11] H. Kagechika, *ISIJ Int.* **47**, 773-794 (2007).
[12] X. M. Wang and X. L. He, *ISIJ Int.* **42**, 38-46 (2002).
[13] Y. Ohmori and K. Yamanaka, in *Boron in Steels* edited by S. K. Banerji and J. E. Morral, (Metall. Soc. AIME, New York (1980) pp. 44-60.
[14] T. Tanaka, *Int. Metals Rev.* **4**, 185-212 (1981).
[15] K. Tsukada, K. Matsumoto, Y. Yamazaki, K. Hirabe, K. Arikata and K. Takeshige, *Tetsu-to-Hagané* **70(1)**, 89-95 (1984).
[16] A. J. DeArdo, *Can. Metall. Q.* **27(2)**, 141-154 (1988).
[17] A. Ghosh, B. Mishra, S. Das and S. Chatterjee, *Metall. Mater. Trans.* **A36**, 703-713 (2005).

Mater. Res. Soc. Symp. Proc. Vol. 1373 © 2012 Materials Research Society
DOI: 10.1557/opl.2012.302

Effect of Hydrogen on Mechanical Properties of Aluminum AA 6061 Alloy Hardened by Precipitation

M. A. Doñu Ruiz[1], V. J. Cortés Suárez[2], N. López Perrusquia[1], L. D. Rosado Cruz[3], F. de J. Valencia Jara[3], A. Altamirano Torres[2]
[1]Instituto Politécnico Nacional,SEPI-ESIME, Adolfo López Mateos, Zacatenco, México D.F., 07738, México.
[2]Universidad Autónoma Metropolitana UAM unidad Azcapotzalco, Departamento de Materiales, Av. San Pablo 180 Azcapotzalco, México .DF, 02200.
[3]Universidad Politécnica del Valle de México UPVM. Av. Mexiquense s/n Esquina Av. Universidad Politécnica, Col Villa Esmeralda, 54910, Tultitlan. Edo de México.

ABSTRACT

The present work reports on the effect of hydrogen on mechanical properties of aluminum AA 6061 hardened by precipitation. Test specimens are solution treated in an electric induction furnace at 803 K during 1 hour and then precipitation hardened during 4 hours at 423, 448 and 473 K. The heat treated specimens are subjected to a hydrogen cathodic charging process during 0, 24 and 96 hours and then tested under monotonic uniaxial tensile loading according to ASTM E8-03. Scanning electron microscopy characterization of tensile fracture surfaces reveals that hydrogen charging of precipitation hardened specimens enhances intergranular cracking by microvoid coalescence and intragranular cleavage fracture. The results show that, although hydrogen charging causes embrittlement of precipitation hardened aluminum AA 6061, the effect is a function of time; short time cathodic charging increases the ductility of the alloy.

INTRODUCTION

Aluminum alloys containing Mg and Si as major alloying additions are widely used for automotive applications because of their good combination of properties. The heat treatable Al-Mg-Si alloys of the 6xxx series are often chosen for these applications, since they show a good combination of formability, corrosion resistance and weldability [2]. The hardening phase in these alloys has recently been identified to be the coherent metastable β'' phase [1]. The introduction of hydrogen into aluminum alloy systems generally has a deleterious effect on the mechanical properties and particularly on the fracture toughness [3, 5]. The embrittlement resulting from the presence of solute hydrogen can be induced during processing or service in a gaseous or liquid corrosive environment. Strengthening of Al-Mg-Si alloys is based on a precipitation hardening process [6–8] and numerous investigations have reported on the effect of external hydrogen intake on the deformation and fracture behavior of precipitation hardened alloys[6,7]. The purpose of this work is to evaluate the effect of cathodic hydrogen charging on the mechanical properties and fracture behavior of aluminum AA 6061 hardened by precipitation.

EXPERIMENT

The nominal chemical composition of the aluminum AA 6061 alloy used in the present study is 0.40-0.80% ***Si***, 0.80-1.20% ***Mg***, 0.7% ***Fe***, 0.15-0.4% ***Cu***, 0.15% ***Mn***, 0.04-0.035% ***Cr***,

0.25% ***Zn***, 0.15% ***Ti***. Tensile specimens are solution treated at 803 K for 1 hour and quenched in water. Artificial aging is carried out at 423, 448 and 473 K during 4 hours followed by water quenching to room temperature. The heat treatments are performed using two Lindberg model 51894 furnaces. Hydrogen cathodic charging is carried out in a 0.5M H_2SO_4 electrolytic solution using a current density of 7.5 mA/cm^2 during 0, 24 and 96 h. Tensile tests are carried out at room temperature according to ASTM E-8-03 using a universal testing machine (UNITED SSTM1). The fracture surfaces of the tested specimens are examined using a scanning electron microscope (SEM JEOL JSM 5900).

RESULTS AND DISCUSSION

Effect of aging heat treatment on mechanical properties

The tensile stress-strain curves for each condition of aging are shown in Figure 1 and Table I lists the tensile properties of AA 6061; ultimate tensile strength (UTS), yield strength (YS), fracture stress (FS), reduction of area at fracture (RA) and elongation to fracture (EL), each data point represents the average of at least five tests.

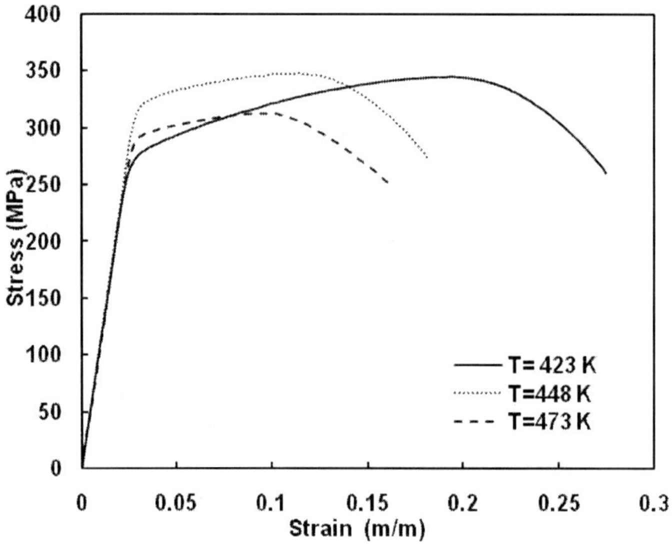

Figure 1. Stress-strain curves for aluminum AA6061 aged at different temperature for 4 hours cooled in water.

Table I. Tensile Properties of AA 6061 with different aging conditions

Temperature (K)	Time (h)	YS (MPa)	UTS (MPa)	FS (MPa)	RA (%)	EL (%)
423	4	266.80	344.80	260.30	35.25	27.80
448	4	311.70	348.00	272.00	34.25	19.70
473	4	288.00	312.30	250.00	35.88	17.20

The optimum strength is observed after aging at 448 K while underaging and overaging are observed at 423 and 473 K, respectively. The same trend is observed for the elongation to fracture with very little effect on the reduction of area at fracture. Thus, according with the results shown in Table I, the optimum aging treatment is 4 hours at 448 K.

Effect of cathodic hydrogen charging.

Figure 2a shows the effects of hydrogen charging on the stress-strain curves of AA 6061 aged 4 hours at 448 K. Figure 2b presents the change in YS, UTS, FS and RA of AA 6061 as a function of charging time and Table II lists the tensile properties for the various conditions tested. The tensile property data shows that hydrogen charging during 24 hours of aged (4 hr, 448 K) AA6061causes a small decrease in YS and UTS. Comparing the data presented in Table I and Table II, it can be observed that a similar effect occurs in underaged and overaged samples. Increasing the charging time to 96 hours causes only a small increase in the tensile strength of the samples (see Table II). In contrast, the ductility of the charged samples, as measured by RA at fracture, first increases significantly after 24 hours and then drops dramatically after 96 hours. In is noteworthy that, on a percentage basis, the effect of hydrogen charging is much more pronounced on RA than on EL. This result is surprising considering that a number of researchers [8-10] have indicated that the ductility of Al-alloys decreases significantly when they are cathodically charged with hydrogen. In the present work, RA exhibits a different behavior for charging times up to 24 hours: RA increases by more than 25% with respect to the uncharged condition (Figure 2b).

Table II. Effect of hydrogen charging on the tensile properties of AA 6061.

Temperature (K)	Charging time (h)	YS (MPa)	UTS (Mpa)	FS (MPa)	RA (%)	EL (%)
423	24	274.5	341.1	277.5	34.00	27.18
448	24	292.4	321.8	229.0	43.25	18.62
	96	290.0	329.0	313.0	21.75	16.00
478	24	280.4	307.0	241.0	31.25	19.93

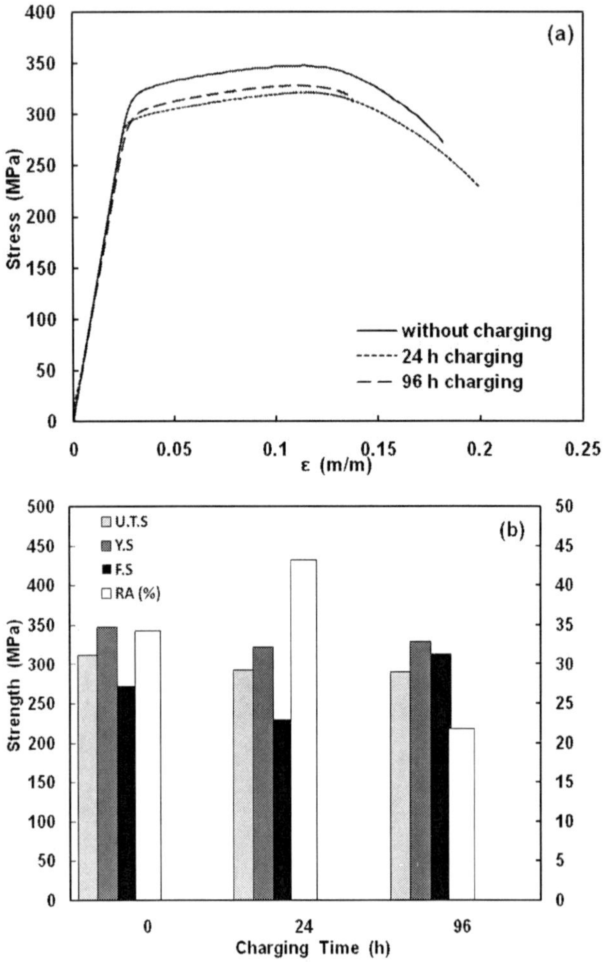

Fgure 2. Effect of hydrogen charging time on: a) stress-strain curves and b) yield strength (YS), ultimate tensile strength (UTS), fracture stress (FS) and reduction of area at fracture (%RA), of AA 6061artificially aged during 4 hours at 448 K.

SEM observation of the fracture surfaces

The effect of hydrogen charging time on the fracture surfaces of specimens of AA 6061 artificially aged at 448 K during for 4 hours is shown in Figures 3 and 4. As can be seen in Fig. 3a, the fracture surface of the specimen aged at 448 K during 4 hours exhibits a fibrous central region, which is associated with the nucleation and growth of the crack, surrounded by a zone of ductile tearing, which represents the final separation resulting in a ductile cup and cone type fracture. Figure 4a illustrates the dimpled morphology (average dimple diameter of 11.6 μm) of the fracture surface which is characteristic of the tensile fracture of high ductility materials. As can be seen, the dimples are separated by thin regions of ductile tearing and exhibit conical shapes with very fine second phase particles within them.

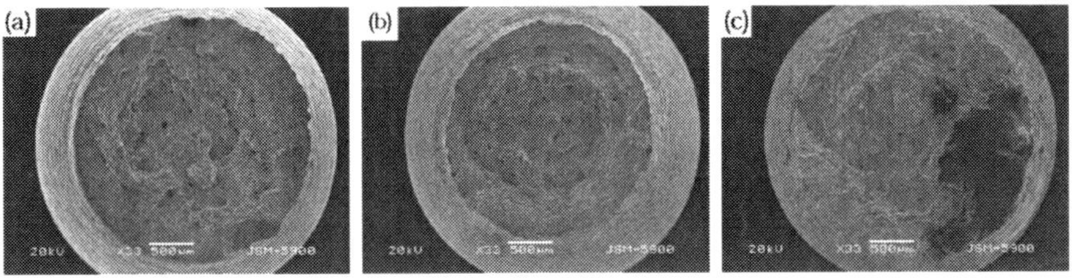

Figure 3. Effect of hydrogen charging time on the fracture surfaces of specimens of AA 6061 artificially aged at 448 K during for 4 hours a) 0 h, b) 24 h, c) 96 h.

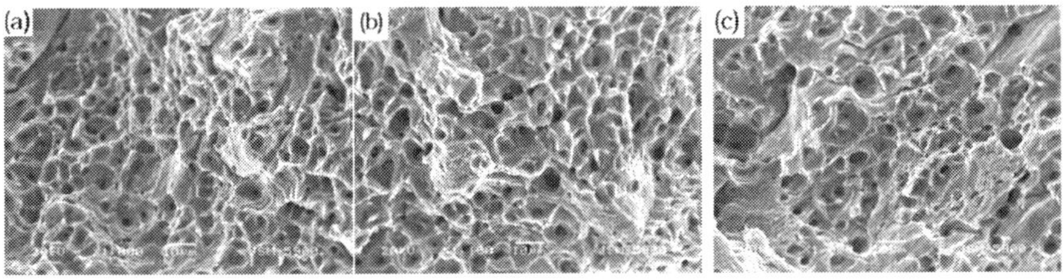

Figure 4. High magnification SEM micrographs of the fracture surfaces of specimens of AA 6061 showing ductile fracture with elongated dimples; (a) charging time: 0 hr, (b) charging time: 24 hr; and mixed ductile-brittle fracture with elongated dimples and intergranular cleavage facets; (c) charging time: 96 hr.

Similar fracture surface characteristics are observed in the sample charged during 24 hours (Figure 3b). In this case, however, there is an increase in the magnitude of the plastic deformation in the center of the specimen which resulted in a larger reduction of area at fracture (see Figure 2b) and an increase in average dimple diameter to 12.4 μm. Some regions of quasi-cleavage fracture can also be observed in Figure 4b: smooth, flat regions separated by zones of ductile tearing. These regions are related to hydrogen induced crack initiation sites. In contrast, the dimples are a clear indication of ductile fracture initiated by nucleation of microvoids at the interfaces between the matrix and the β (Mg_2Si) precipitate particles.

As can be appreciated in Figures 3c and 4c, hydrogen charging during 96 hours significantly changed the mode of fracture compared with the uncharged and 24-hour charged specimens. In this case, the cup-cone type of fracture is not developed and the reduction of area at fracture is significantly reduced (Figure 3c). In addition, at high magnifications (Figure 4c), intergranular brittle fracture regions are clearly visible and the dimple diameter and density in the central part of the fracture surface are severely reduced. Also, regions of dimpled ductile fracture are separated by large regions of transgranular fracture where faceted patterns and steps suggest the occurrence of cleavage.

CONCLUSIONS

Optimum mechanical properties of aluminum AA 6061 alloy are obtained after artificial again during 4 hours at 448 K. Hydrogen cathodic charging during 96 hours of aged specimens causes embrittlement. However, the results of the present investigations showed that short-term (up to 24 hours) cathodic hydrogen causes a 25% increase in reduction of area at fracture. Increasing the hydrogen charging time from 24 to 96 hours changes the appearance of the fracture surface from 100% dimpled cup-cone to mixed ductile-brittle fracture with elongated dimples and intergranular cleavage facets.

ACKNOWLEDGMENTS

Two of the authors gratefully acknowledge the support given by *Consejo Nacional de Ciencia y Tecnologia,* CONACyT through the Doctoral Program of Instituto Politecnico Nacional

REFERENCES

1. G. A. Edwards, k. Stiller, g. L. Dunlop and m. J. Couper, Acta Metallurgica 46 (1998) 3893.
2. G. B. Burger, A. K. Gupta, P. W. Jeffrey, and D. J. Lloyd, Mater. Charact. 35 (1995) 23.
3. L. Cristodoulou, H. M. Flower, Acta Metallurgica, 28 (1980) 481-487.
4. A. I. Onuchukwu, S.P. Trasatti, S. Trasatti, Corrosion Science, 36 (1994) 1815.
5. Wookil Jang, Sangshik Kim, Kwangseon Shin. Metallurgical and Materials Transactions, 33 (2002) 1755.
6. R. P. Wahi and M. von Heimendahl, Phys. Stat. Sol. (a). 24, (1974) 607.
7. S. J. Andersen, H. W. Zandbergen, J. Jansen, C. Treholt, U. Tundal, and O. Reiso, Acta Mater. 46, 3283 (1998).
8. ASM Handbook, Volume 4, *Heat Treating.*1997.pg. 1862-1960.
9. Jang, W. K. and Kim, S. S., Scripta Materialia 67 (1999) 503.
10. Wookil Jang, Sangshik Kim, Kwangseon Shin, Metallurgical and Materials Transactions A, 33a (2002) 1757.

Mater. Res. Soc. Symp. Proc. Vol. 1373 © 2012 Materials Research Society
DOI: 10.1557/opl.2012.303

Numerical Processing of Cooling Curves to Obtain Growth Parameters During Eutectic Solidification

M. Morua[1], E. Peña[1], R. Aparicio[2], M.Ramirez-Argaez[1] and C. Gonzalez-Rivera[1]
[1]Departamento de Ingeniería Metalúrgica, Facultad de Química, UNAM, Edificio D Circuito de los Institutos s/n Cd. Universitaria Mexico D.F., Mexico.
[2]Cinvestav Querétaro Libramiento Norponiente 2000, Real de Juriquilla, 76230 Querétaro Qro. Mexico.

ABSTRACT

In this work a methodology to measure kinetic parameters to describe grain growth during equiaxed eutectic solidification is proposed. This methodology includes the numerical processing of two cooling curves and requires input data concerning the number of grains per unit volume. In addition, free grain growth before impingement and an exponential dependence of the grain growth rate on undercooling are assumed. The evolution of solid fraction of the sample as a function of time is obtained by applying the Fourier thermal analysis (FTA) method. Information collected is processed numerically in order to find numerical values for the pre-exponential and exponential parameters that characterize the grain growth kinetics as a function of undercooling. To validate this methodology a mechanistic model that simulates the cooling and solidification of eutectic Al-Si alloy in a sand mold is used. The results suggest that this methodology can be used to measure the kinetic parameters of equiaxed growth from the numerical processing of cooling curves and grain density data.

INTRODUCTION

The control of solidification microstructures is vital for the development and improvement of the quality and properties of the final cast products. Prediction of the formation and evolution of the cast microstructure is an essential element in the control of these microstructures during solidification and therefore the modeling of these microstructures is of considerable interest. As a result, several kinds of models have been developed in the last couple of decades [1-4]. These models include calculations on nucleation and growth kinetics which in turn depend on the availability of nucleation and growth models capable to reproduce the experimental behavior of the alloy under study. This work is focused on the development of a methodology which allows obtaining growth models for equiaxed eutectic solidification from experimental data.

It has been found that, during solidification, grain growth rate depends on undercooling, defined as the difference between equilibrium solidification temperature and the actual temperature present during the phase change. Previous efforts have been done to develop methodologies allowing grain growth characterization [5, 6]. The following exponential equation has been used to describe grain growth, where R is the mean grain radius, t is the time, dR/dt is the grain growth rate, ΔT is the undercooling, and μ and n are the grain growth coefficients.

$$\left(\frac{dR}{dt}\right)=\mu\Delta T^{n} \quad (1)$$

The purpose of this work is the development and validation of a method for determination of equiaxed eutectic grain growth kinetic coefficients using the cooling curve data associated with the initial stage of growth after nucleation. A simplified free equiaxed growth model, equiaxed grain density measurements and the Fourier thermal analysis method (FTA) to obtain the solid fraction evolution are used to follow the grain growth kinetics associated to a cylindrical probe unidirectionally solidified in the radial direction.

Free growth model and method of determination of growth coefficients.

In order to obtain the growth coefficients from experimental data it is assumed that Eq. (1) describes the dependence of the grain growth rate on undercooling. Thus, a plot of log(dR/dt) versus log(ΔT) , would give values of n and μ according to equation (2):

$$\log\left(\frac{dR}{dt}\right)=\log(\mu)+n\log(\Delta T) \qquad (2)$$

The growth rate dR/dt can be obtained using a free grain growth model and known data on solid fraction evolution and grain density. In this way, an equivalent grain radius can be obtained every time step recorded in the cooling curve.

It is assumed that during the solidification of the sample there are N spherical grains of radius R developing at the same time. N is the number of grains per unit volume, so that the solid fraction in the early stages of growth at a time t is given by the following equation:

$$Fs^t=\frac{4}{3}\pi N\left(R^t\right)^3 \qquad (3)$$

Fs^t and R^t are the solid fraction and grain radius at time t, respectively. The density of grains per unit volume in the sample, N, can be determined by metallographic methods and Fs is obtained from the Fourier Method applied to cooling curves at two radial positions in the sample. Thus, it is possible to estimate the instantaneous grain radius using Eq. (4):

$$R^t=\left(\frac{3Fs^t}{4\pi N}\right)^{\frac{1}{3}} \qquad (4)$$

The evolution of grain radius as a function of time can be obtained from the following expression:

$$\frac{dR}{dt}=\frac{R^{t+dt}-R^t}{\Delta t} \qquad (5)$$

The instantaneous undercooling during solidification can be obtained directly from the cooling curve at any time using the following equation:

$$\Delta T=T_{Eu}-T \qquad (6)$$

T_{Eu} is the equilibrium eutectic temperature and T is the instantaneous temperature recorded by the thermocouple during solidification.

Heat transfer and solidification kinetics model

In order to validate the predictions of the free growth method and the values obtained for the kinetic coefficients, the proposed method is tested using cooling curves and grain density information provided by a heat transfer solidification kinetics (HT/SK) model previously reported [7].The model assumes that during the cooling and solidification of a cylindrical casting in a sand mold the macroscopic heat flow is governed by conductive heat transfer and latent heat generation due to solidification. It is also assumed that the thermal conductivity is constant and that the heat fluxes in the axial and azimutal directions of the cylindrical system are zero (unidirectional radial heat flow). Thus, the energy balance applied to the metal/mold system can be written as:

$$Cp_j^v \frac{\partial T(r,t)}{\partial t} = k_j^{th} \frac{1}{r}\frac{\partial}{\partial r}\left(r \frac{\partial T(r,t)}{\partial r} \right) + L_f^v \frac{\partial fs(r,t)}{\partial t} \tag{7}$$

where Cp is the volumetric heat capacity, T(r, t) is the temperature, r is the radial position, t is the time, k^{th} is the thermal conductivity, L_f^v is the volumetric heat of fusion and the subscript j indicates the metal domain (j=1) or the mold domain (j=2). The model also assumes instantaneous nucleation of the equiaxed eutectic grains and uses an empirical parabolic nucleation law that correlates the number of eutectic nuclei with the cooling rate:

$$N = A + B\left(\frac{dT}{dt} \right)^2_{T=T_{Eu}} \tag{8}$$

where N is the density of grains per unit volume and the parameters A and B are determined experimentally from grain density versus cooling rate data. The eutectic grains are assumed to be spherical in shape and the rate of growth of eutectic grains is calculated using Eq. (1). Table 1 gives the nucleation and growth parameters used during the calculations.

Table I. Nucleation and growth parameters used during simulation

A=5.21×10^8	m^{-3}
B=3.56×10^6	$m^{-3}\ S^2\ {}^oC^{-2}$
$\mu = 5 \times 10^{-6}$	$m\ S^{-1}\ {}^oC^{-2}$
n = 2	-

Eq. (7) is solved by generating a discretized description of the cylindrical metal/mold system in the form of a finite difference mesh composed by a known number of cylindrical volume elements. The assumptions, including initial and boundary conditions, used to solve Eq. (7) are described in a previous work [7].

Fourier thermal analysis method

The FTA method has been described in detail elsewhere [8]. Briefly, two cooling curves obtained at two radial locations within a cylindrical sample with thermally isolated top and bottom are numerically processed. The FTA numerical processing starts with the generation of the first derivative with respect to time of the curve corresponding to the thermal center of the sample and the identification of the times of start and end of solidification. FTA uses the data acquired from the two thermocouples to obtain the thermal diffusivity of the sample and the zero baseline curve by an iterative procedure. The integration of the area between the first derivative of the cooling curve and the zero baseline curve gives relevant quantitative information on solidification kinetics. The fundamentals, limitations, and implementation of this method have been discussed elsewhere [9-11]

RESULTS AND DISCUSSION

The HT/SK model is used to simulate the cooling and solidification of a cylindrical casting of eutectic Al-Si alloy in a sand mold. The radius of the casting is r=0.03 m and two cooling curves are obtained for volume elements located a two radial positions (r_1= 0 and r_2=0.005m). The cooling curves are simulated by the model and numerically processed using the FTA method. The density of grains per unit volume predicted by the model is $5.2x10^8$ grains/m^3.

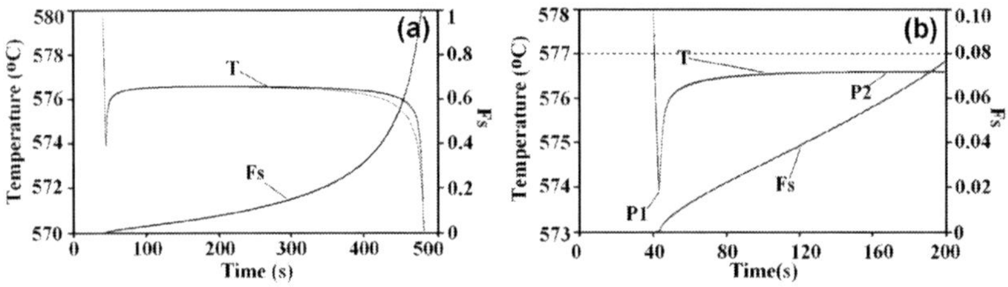

Figure 1. Cooling curves predicted by the model and solid fraction evolution obtained from numerical processing using FTA.; **(a)** Model predictions and FTA solid fraction evolution during solidification. ; **(b)** Detail of curves during the time interval used to perform grain growth characterization.

Figure 1a shows the cooling curves predicted by the model and the solid fraction evolution obtained from FTA numerical processing of these curves. As can be seen in the cooling curves, there is a first stage of cooling of the liquid indicated by the continuous drop in temperature from the beginning of cooling. As the curve approaches the eutectic temperature, this initial stage is followed by a plateau corresponding to the solidification of the eutectic microconstituent.

Finally, after the eutectic plateau, the temperature drops again during the final stage of cooling of the solid alloy.

The Figure 1b shows in more detail the stage of eutectic solidification and the parts of the cooling curve and the solid fraction curve that are used for grain growth kinetic calculations.

The cooling curve data used to calculate the thermal undercooling, defined as the difference between equilibrium Al-Si eutectic temperature (577°C) and the temperature measured during eutectic growth, according to Eq.(6), includes the section of the curve corresponding to the point of maximum undercooling and the point of maximum recalescence, shown as points P1 and P2 in this Figure 1b. During this time interval, using the known number of grains per unit volume and the solid fraction evolution as a function of time predicted by the Fourier Method, it is possible to calculate the evolution of the grain radius. Figure 2 shows the plot of log(dR/dt) versus log(ΔT) generated by the free growth method.

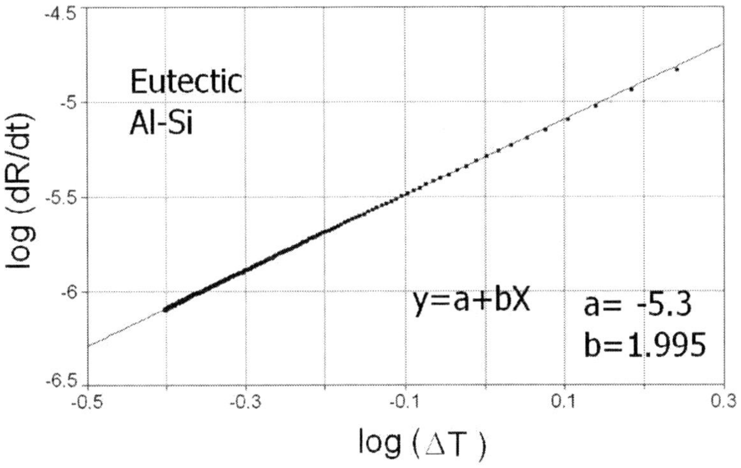

Figure 2. Logarithmic plot of grain growth rate against undercooling obtained from free growth method.

As can be seen, a good fit to a linear equation, with an intercept of -5.3 , i.e. a pre-exponential growth coefficient of μ=5.01 x 10^{-6} , and an exponential coefficient of 1.995 are obtained. Comparing these results with the growth coefficients used in the model (μ=5 x 10^{-6} and n=2, see Table I), it can be concluded that the proposed methodology can be used to obtain the growth coefficients for equiaxed eutectic grain development.

CONCLUSION

A new methodology for the determination of equiaxed eutectic grain growth kinetic coefficients is proposed in this work. This methodology requires knowledge of cooling curve data associated to the initial stages of growth after nucleation, the eutectic grain density, a simplified free equiaxed growth model and the Fourier thermal analysis method of the cooling curves.

The method has been validated using predicted cooling curves generated by a heat transfer/solidification kinetics model intended to simulate the cooling and solidification of an eutectic Al-Si alloy in a sand mold.

The results indicate that the proposed methodology can be used successfully to obtain the growth coefficients for equiaxed eutectic grain development.

ACKNOWLEDGEMENTS

The authors acknowledges DGAPA UNAM, for its financial support (Project IN112209).and to A. Ruiz, A. Amaro and C. Atlatenco for their valuable technical assistance
C. Gonzalez-Rivera acknowledge CONACYT Mexico for its support through a sabbatical grant.

REFERENCES

1. B. Liu , Q. Xu, T. Jing, H. Shen and Z. Han. *J. of Metals*, 63, 4, 19,(2011).
2. M.F. Zhu , C.P. Hong and D.M .Stefanescu . *Met. Mater. Trans B*, 38B, 517,(2007).
3. X. Yao, M.S. Dargusch and A.K. Dahle, *J. Mater. Res.*, 23,9, 2312 (2008).
4.H. Zhang, K. Oikawa, M. Ohno and Pär G. Jönsson. *ISIJ International, 50 ,12, 1724, (2010).*
5. Dioszegi A. and Svensson I. L *Int.J. of Cast Metals Research*, 18 , 1, 41(2005).
6 Degand C., D.M. Stefanescu and G. Laslaz in: "Solidification Science and Processing", I. Ohnaka and D.M. Stefanescu eds., TMS, Warrendale Pa., 55 (1996).
7. C.González-Rivera, H. Cruz M.,A.García H. J.Juárez-Islas. *J. of Mater. Eng. and Performance* 8,1,103,(1999).
8. W. Kapturkiewicz, A. Burbielko, and H.F. Lopez, *AFS Trans.*, 101, 505, (1993).
9. J.O. Barlow and D.M. Stefanescu, *AFS Trans.,* 105, 339, (1997).
10. E. Fras, W. Kapturkiewicz, A. Burbielko, and H.F. Lopez, *Metall. Mater. Trans. B*, 28B, 115, (1997).
11. D. Emadi and L. Whiting, *AFS Trans.,* 110,285, (2002)

Mater. Res. Soc. Symp. Proc. Vol. 1373 © 2012 Materials Research Society
DOI: 10.1557/opl.2012.304

Sheet moulding compounds with improved mechanical properties

V. Nicolau[1,2], N. Bertoneri[1], J. Scarello[1], D. Estenoz[2]
[1]Dpto. Ingeniería Química, Fac. Reg. San Fco. (U.T.N.), Av. de la Universidad 501, San Francisco (2400), Córdoba, Argentina.
[2]INTEC (CONICET and U.N.L.), Güemes 3450, Santa Fe (3000), Argentina.

ABSTRACT

This work studies the synthesis of styrene copolymers obtained from the reaction between isophthalic unsaturated polyester resins and styrene with the aim of improving the mechanical properties of sheet moulding compounds (SMC), specifically their flexibility and impact resistance.
The experimental work involves the experimental design; the SMC synthesis with different concentrations of 4 additives (methylmethacrylate; diethylenglycol, dioctilphthalate, and a commercial flexible resin); and the mechanical characterization including measurements of load-deflection curves; flexural strength (FS), flexural modulus (FM) and impact resistance (IR).

INTRODUCTION

SMC are used extensively in the production of thermosetting composite materials for use in the automotive, trucking, construction, civil infrastructure, aircraft, and military industries. The moulding compound consists of a mixture of a liquid moulding resin, fibres and/or fillers, and additives. The most common resins used in SMC are the various unsaturated polyester and vinyl ester resins. When reinforced with materials such as glass fibres, they can be formed into strong, light laminates used to produce boats, swimming pools, spa baths, and translucent roof sheet. These resins can also be filled with calcium carbonate and/or alumina trihydrate powders and cast into items such as synthetic vanity basins and bench tops. However, they are typically manufactured in sheet form for fabrication into finished countertops known as "solid surface". Solid surface is a non-porous low-maintenance countertop surface. It can mimic the appearance of stone, marble, onyx and granite and other naturally occurring materials, and can be joined invisibly by a trained craftsman.

These copolymers are obtained from radical reaction between styrene (cross-linking agent) and an unsaturated polyester prepolymer (orthophthalic, isophthalic or flexible) or a vinyl ester prepolymer. The latter are more flexible but also more expensive.

Basically, unsaturated polyester prepolymers are obtained from a condensation reaction between a saturated dibasic acid $HOOCR_1COOH$, maleic anhydride $C_4H_2O_3$ and an aliphatic diol HOR_2OH.

Orthophthalic, isophthalic, and adipic (saturated) acids are employed for the synthesis of orthophthalic, isophthalic and flexible prepolymers, respectively. The saturated acid decreases the crosslink density during the curing stage. Orthophthalic resins are more brittle than isophthalic resins, and flexible resins are employed together with orthophthalic and isophthalic resins in order to increase their flexibility. These prepolymers are sold as "unsaturated polyester resins" (UP-resins). The resin consists of linear unsaturated oligomers dissolved in styrene monomer (Figure 1a). The styrene acts as a solvent and also as the monomer that reacts with the double bonds of the polyester as it polymerizes. The reaction is usually initiated by the addition of a peroxide (R-R). The curing agent R-R splits into two radicals on heating or under UV-light or on the addition of a promoter (accelerator) which allows room temperature curing. These radicals then initiate the polymerization of the styrene that results in cross-linking of the UP-resins through short polystyrene chains (Figure 1b).

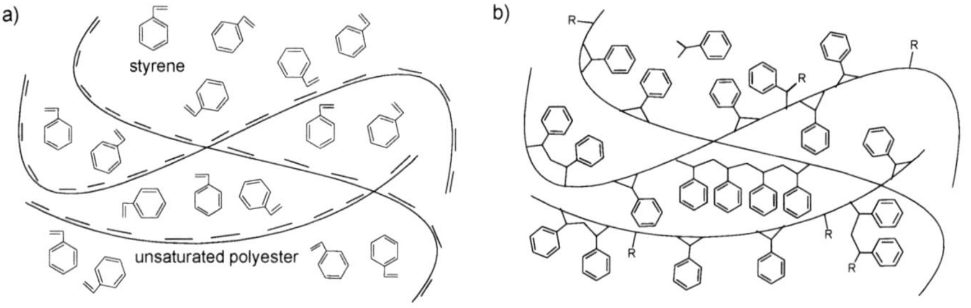

Figure 1. a) UP-Resin, b) Styrene Copolymer.

This work studies the synthesis of styrene copolymers based on isophthalic UP-resins for the production of SMC known as "solid surface" in order to improve their mechanical properties, in particular flexibility and impact resistance. The aim of this work is the optimization of the industrial process of the only firm in the country that produces this material (Tresol s.r.l, San Francisco, Córdoba, Argentina).

EXPERIMENTAL WORK

The materials are synthesized from a base mixture prepared by the addition of isophthalic UP-resin, the promoter, and alumina trihydrate and dolomite as fillers in a mass fillers:resin ratio ≅ 2:1. The base mixture is fractionated into moulds with the addition of the respective additives (methylmethacrylate, MMA; diethylenglycol, DEG; dioctilphthalate, DOP; and a commercial flexible resin, P1300 Poliresinas, San Luis, Argentina) in ml/100g mixture, and methyl ethyl ketone as initiator [1]. Air bubbles are eliminated in a vibrating table during 30 min. and samples are cured at 80-90 °C during 2-3 hours. Specimens for flexural measurements and impact resistance are obtained from the samples. Recipe and operative conditions during the production of the SMC samples are shown in Table I.

Table I . Recipe and operative conditions during the production of the SMC samples.

Recipe:	Parts (w)
Isophtalic resin	100
Alumina trihydrate	100
Dolomite	100
Methyl ethyl ketone	1
Cobalt octoacte	0.3
Mixture Conditions:	
Temperature, °C	22 ± 0.5
% Relative humidity	30
Time, min.	40

Flexural tests are performed in a Shimadzu, model DSS-10T, at a rate of 5.0 mm/min. according to IRAM 13.338 standard [2]. Three specimens per sample are cut according to the dimensions specified by the standard (8.0 x 1.0 x 0.4 cm). Figure 2 shows the load-deflection diagrams up to the moment of rupture. Flexural Strength (FS) and flexural modulus (FM) shown in Table II are obtained from these diagrams.

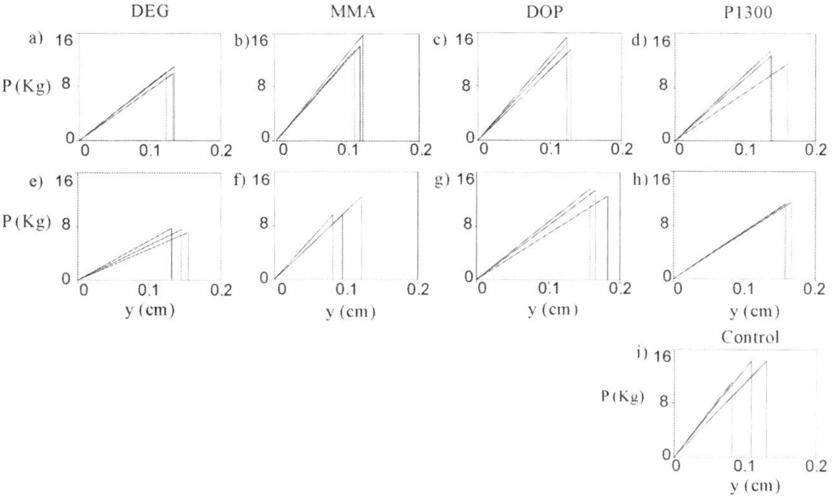

Figure 2. Load-Deflection diagrams for materials containing additives in 2 (a-d) and 5 (e-h) ml/100g mixture; and the control (i).

The impact tests are accomplished according to method A (Izod with notch) established by the ASTM D 256:10 standard, in a Custom Scientific Instrument, model CS 137 [3]. Ten rectangular specimens with notch for sample are tested. Specimen dimensions and impact resistance (IR) measurements are shown in Table III. Note that in some cases the number of repetitions are lower than ten due to the lack of samples.

Table II. Flexural resistance (FR) and flexural modulus (FM) for the control and samples with 4 kinds of additves (DEG, MMA, DOP y P1300). Average values are in bold font.

ml/100g mixture	Additive	Specimen Width (b) [cm]	Specimen Height (h) [cm]	Diagram P [kg]	Diagram y [cm]	FS [daN/cm²]		FM [daN/cm²]	
0	Without additive (control)	1.01	0.42	14.4	0.140	750		81737	
		1.03	0.42	14.0	0.120	719	**736**	91446	**84576**
		1.03	0.42	14.4	0.140	739		80544	
2	DEG	1.05	0.43	10.0	0.145	479		49221	
		1.01	0.47	11.2	0.148	467	**463**	43145	**46050**
		1.04	0.46	10.4	0.135	444		45783	
	MMA	1.03	0.42	13.6	0.123	703		87533	
		1.02	0.42	14.4	0.130	752	**761**	88199	**89725**
		1.02	0.42	16.0	0.135	827		93444	
	DOP	1.01	0.42	15.6	0.135	814		92010	
		1.02	0.41	14.4	0.135	781	**778**	90405	**87768**
		1.02	0.41	13.6	0,143	738		80889	
	P1300	1.06	0.42	11.6	0.148	580		59949	
		1.05	0.42	13.6	0.148	685	**636**	70822	**65376**
		1.05	0.42	12.8	0.150	643		65358	
5	DEG	0.96	0.40	7.6	0.160	460		46064	
		1.02	0.47	7.8	0.145	322	**385**	30266	**36574**
		1.03	0.42	7.2	0.170	372		33393	
	MMA	1.01	0.43	9.6	0.105	478		67838	
		1.01	0.44	9.6	0.090	457	**508**	73869	**68839**
		1.02	0.44	12.4	0.133	590		64810	
	DOP	1.03	0.42	13.6	0.183	696		58184	
		1.02	0.43	13.8	0.173	680	**693**	58776	**56858**
		1.03	0.42	13.6	0.200	703		53614	
	P1300	1.01	0.41	11.2	0.173	613		55519	
		1.01	0.41	11.4	0.175	623	**618**	55648	**54718**
		1.04	0.41	11.6	0.183	619		52988	

DISCUSSION

A two-factor (additive and concentration) completely random experimental design is used for each of the 3 variables (FS, FM, and IR) with 4 levels of additives (MMA, DEG, DOP, and P1300) and with 2 levels in compositions (2 and 5 ml/100g mixture). The design is analyzed using Analysis of Variance (ANOVA) employing Statgraphics Centurion XV software [4]. From the Anova analysis it can be seen that both the kind of additive and its concentration have a statistically significant effect on FS and FM.

Table III. Impact Resistance (IR) for the control and samples with 4 kinds of additives (DEG, MMA, DOP y P1300). Average values are in bold font.

ml/100g mixture	Additive	Specimen Width [mm]		Specimen Height [mm]		Energy [J]		IR [KJ/m^2]		
		4.36	4.43	12.35	12.24	0.071	0.067	1.57	1.48	
	Without	4.38	4.43	12.22	12.37	0.060	0.075	1.34	1.63	
0	additive	4.43	4.36	12.36	12.46	0.067	0.075	1.46	1.64	**1.54**
	(control)	4.43	4.43	12.38	12.75	0.075	0.075	1.63	1.57	
		4.43	4.35	12.50	12.29	0.064	0.075	1.38	1.68	
		4.17	4.17	12.55	12.50	0.078	0.078	1.77	1.78	
	DEG	4.18	4.16	12.41	12.35	0.070	0.066	1.61	1.53	**1.64**
		4.15	4.19	12.44	12.31	0.082	0.060	1.89	1.39	
		4.19	4.16	12.23	12.19	0.074	0.060	1.73	1.42	
		4.05	4.12	12.54	12.41	0.071	0.075	1.66	1.75	
	MMA	4.05	4.13	12.26	12.83	0.067	0.079	1.61	1.77	**1.74**
		4.22	4.19	12.16	12.38	0.060	0.087	1.40	2.00	
2		4.04	4.21	12.31	12.25	0.067	0.079	1.61	1.83	
		4.08	---*	12.23	---*	0.083	---*	1.99	---*	
		4.23	4.33	12.02	12.24	0.126	0.064	2.97	1.44	
	DOP	4.34	4.29	12.16	12.24	0.103	0.06	2.34	1.37	**2.21**
		4.31	4.36	12.31	12.19	0.169	0.067	3.80	1.51	
		4.36	---*	12.35	---*	0.091	---*	2.02	---*	
		4.34	4.33	12.45	12.48	0.189	0.083	4.17	1.65	
	P1300	4.39	4.33	12.67	12.51	0.095	0.075	2.03	1.56	**2.09**
		4.38	4.21	12.46	12.65	0.107	0.071	2.34	1.85	
		4.28	4.37	12.45	12.70	0.079	0.083	1.77	1.52	
		4.31	---*	12.21	---*	0.083	---*	1.89	---	
		4.26	4.31	12.69	12.36	0.067	0.067	1.47	1.50	
	DEG	4.25	4.27	12.30	12.11	0.060	0.075	1.37	1.74	**1.49**
		4.27	4.32	12.57	12.34	0.071	0.052	1.57	1.16	
		4.29	4.30	12.28	12.31	0.064	0.060	1.45	1.35	
		4.28	4.29	12.13	12.39	0.071	0.071	1.64	1.59	
		4.10	4.09	12.44	12.52	0.079	0.071	1.85	1.65	
	MMA	4.05	3.98	12.24	12.59	0.071	0.071	1.71	1.68	**1.79**
		4.09	4.07	12.43	12.32	0.071	0.087	1.66	2.07	
5		4.09	3.96	12.37	12.26	0.071	0.079	1.67	1.94	
		4.02	---*	12.26	---*	0.075	---*	1.82	---*	
		4.22	4.37	11.35	12.18	0.067	0.075	1.70	1.69	
	DOP	4.39	4.36	12.19	12.25	0.103	0.075	2.30	1.68	**1.84**
		4.34	4.34	12.25	12.11	0.060	0.083	1.35	1.89	
		4.38	4.35	12.27	12.30	0.083	0.103	1.85	2.30	
		4.19	4.14	12.58	12.34	0.067	0.067	1.51	1.57	
	P1300	4.14	4.17	12.36	12.54	0.075	0.079	1.75	1.80	**1.74**
		4.11	4.23	12.22	12.23	0.079	0.071	1.88	1.64	
		4.18	4.16	12.59	12.32	0.087	0.081	1.97	1.89	
		4.20	4.20	12.25	12.51	0.071	0.079	1.65	1.79	

* Measurements not taken due to the lack of samples.

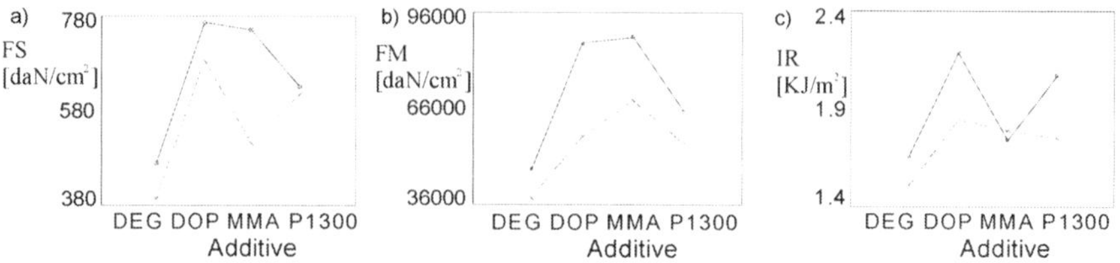

Figure 3. Interaction Plots for FS (a), FM (b), and IR (c). full line: 2 (ml/100g mixture), dashed line: 5 (ml/100g mixture).

Both FS and FM increase in the following additive order: DEG<(P1300=MMA)<DOP; DEG<P1300<DOP<MMA, respectively; and in the following concentration order: 2<5. Figures 3a and 3b show that there is not interaction between factors (parallel lines). Both FS and FM always decrease while increasing the concentration of the additive. Concentration has not a statistically significant effect on IR and there is significant interaction between factors as shown in Figure 3c) (not parallel lines). An increase in MMA concentration shows an increase in IR.

In addition, one way Anova analysis is carried out for each additive and for 3 levels of compositions (0, 2 y 5 ml/100g mixture). The addition of DOP does not affect FS when compared with the control. Lower values of FS are observed for DEG, P1300 and MMA, except for 2 ml MMA/100g mixture. FM also decreases without important changes for MMA and DOP (2 ml /100g mixture). Only MMA has a statistically significant increase in IR when compared with the control. The addition of DOP and P1300 (2 ml/100g mixture) increase IR but this effect disappears for a higher concentration.

CONCLUSIONS

The addition of 5 ml dioctilphthalate /100g mixture improved the flexibility of the SMC with a decrease of flexural modulus without important changes in flexural strength and impact resistance. The impact resistance of SMC increased with the addition of methylmethacrylate. The addition of 5 ml methylmethacrylate/100g mixture improved both flexural modulus and impact resistance but decreased the flexural strength of SMC. Future work will address the influence of additives on the structure-property relationships of SMC.

REFERENCES

1. F. Djaider, *J. Thermop. Comp. Mat.* **20**(1), 53-64 (2007).
2. Ensayo de Flexión, IRAM 13338.
3. Standard Test Methods for Determining the Pendulum Impact Resistance of Notched Specimens of Plastics; ASTM D 256:10, ASTM, PA.
4. D. C. Montgomery, "Diseño y Análisis de Experimentos" (Limusa-Wiley, 2002) pp. 85-118 and 175-228.

Mater. Res. Soc. Symp. Proc. Vol. 1373 © 2012 Materials Research Society
DOI: 10.1557/opl.2012.305

Effects of Al_3Ti and $(Al,Ni)_3Ti$ Addition on the Sintering of TiB_2

Masashi Yoshida
Ube National College of Technology, 2-14-1 Tokiwadai, Ube, Yamaguchi 755-8555, Japan

ABSTRACT

Titanium diboride has been sintered using spark plasma sintering (SPS). With the addition of tetragonal Al_3Ti or cubic $(Al,Ni)_3Ti$, the Vickers hardness, Hv, of TiB_2 increases up to as high as 2000 by the sintering at 1273K, while that of the sample sintered without additives is as low as 20. Such a remarkable improvement is caused by the improvement of direct link between TiB_2 grains in addition to the effect that Al_3Ti or $(Al,Ni)_3Ti$ fills in the space between TiB_2 grains and act as a binder.

INTRODUCTION

Titanium diboride has many desirable properties such as high hardness (Hv=3400), high melting point (3000K), low density, high electrical conductivity and good corrosion resistance [1]. It is a promising candidate for aircraft and automotive application, space vehicle thermal protection and so on [2]. However, a high sintering temperature above 2000K, is required to obtain TiB_2 of high quality, which limits its application [2-5].

Titanium trialuminide (Al_3Ti) has low density, high melting temperature (1600K), good high-temperature strength and good oxidation resistance [6-8], so it is also a promising candidate for aerospace and automotive applications. Tetragonal $D0_{22}$ structure of Al_3Ti can be changed to cubic $L1_0$ structure by substituting some (~10%) of the Al by transition metals such as Ni, Cr and Fe.[9]

Fu et al have calculated the adhesion work among TiB_2, Al and Al_3Ti and found a good binding between TiB_2 and Al_3Ti [10]. In the present study, TiB_2-Al_3Ti and TiB_2-$(Al,Ni)_3Ti$ composites are fabricated using spark plasma sintering at 1273K. Structures and mechanical properties of the composites have been investigated. It has been shown that TiB_2 and Al_3Ti co-exist in equilibrium at 1273K. High Vickers hardness, Hv, of around 2000 has been obtained for the composition of Al_3Ti or $(Al,Ni)_3Ti$ of 30wt%.

EXPERIMENT

The raw materials were commercially available TiB_2 powders of 99.9% purity, Ti powders of 99.5% purity Ni powders of 99.9% purity and Al powders of 99.9% purity (all powders by Furuuchi Chemical Corp.). The diameter of TiB_2 powder particles was several μm and those of Al Ni and Ti powders were around 10 μm. To obtain fine Al_3Ti and $(Al,Ni)_3Ti$ powders, Ti powders were hydrogenated by heating at 873K during 1 hour in H_2 environment. Then, powders of TiHx, Al and Ni were mixed, pulverized and heated at 973K during 1 hour in Ar environment. By this procedure, fine powders of Al_3Ti or $(Al,Ni)_3Ti$ of several μm in diameter could be obtained.

The weight ratio of Al_3Ti in the starting materials were chosen as 0% (pure TiB_2), 10%, 20%, 30%, 40%, 50% and 100% (pure Al_3Ti). The weight ratio of $(Al,Ni)_3Ti$ in the starting materials were chosen as 20% and 30% where the mole ratio of Ni:Al is chosen as 15%. A graphite die, having an internal diameter of 20mm and a wall thickness of 10mm was filled with 5g of the mixed powders, sealed by two graphite punches and mounted on the equipment, SPS-1050 fabricated by Sumitomo Coal Mining Ltd. The mixed powders were heated to 1273K at a rate of 100K/min, kept at 1273K during 10 minutes, and then cooled to room temperature. The temperature was measured using a thermocouple attached to the graphite die. The real temperature in the powder compact may be 50 degrees higher than that of the graphite die [11]. Sintering was performed in vacuum with a residual pressure of 50Pa. A uniaxial pressure of 20MPa was applied during sintering.

To examine the cross-section, the specimens were cut by using a diamond cutter and etched with a solution of HNO_3. SEM observation was performed using a Hitachi H-3300 instrument. X-ray diffraction patterns were measured in a Rigaku X-ray diffractmeter with Cu-kα radiation source. Vickers hardness was measured using a Shimadzu HMV-2 micro hardness tester with a load of 1.96N and pressing time of 15 seconds.

RESULTS AND DISCUSSION

Figure 1 shows the X-ray diffraction patterns of the cross-section of the sintered TiB_2-30%Al_3Ti (Figure 1a) and TiB_2-30%$(Al,Ni)_3Ti$ (Figure 1b) composites. As can be seen, Figure 1a shows the diffraction peaks from hexagonal TiB_2 and tetragonal Al_3Ti while Figure 1b shows diffraction peaks corresponding to hexagonal TiB_2 and cubic $(Al,Ni)_3Ti$. These results show that TiB_2 and Al_3Ti and/or $(Al,Ni)_3Ti$ co-exist in equilibrium at 1273K.

Figure 2 shows the SEM image of the cross-section of the sintered TiB_2. The TiB_2 specimen sintered at 1273K is granular. The diameter of TiB_2 grains is the same as that of the starting TiB_2 powders. The link between grains seems to be loose. Figure 3 shows the SEM images of the TiB_2-30%Al_3Ti (a) and TiB_2:30%$(Al,Ni)_3Ti$ (b) composites. In Figures 3a and 3b, it is found that Al_3Ti and/or $(Al,Ni)_3Ti$ fills the space between TiB_2 grains and dense samples are obtained.

Figure 4 shows the variation of the relative densities of TiB_2-Al_3Ti and TiB_2-$(Al,Ni)_3Ti$ composites sintered at 1273K. The relative density of TiB_2 sintered at 1273K is as low as 50%. It increases as the amount of Al_3Ti increases and a relative density greater than 90% is obtained for the sample with more than 30wt% Al_3Ti. The density of TiB_2-$(Al,Ni)_3Ti$ is higher than that of TiB_2-Al_3Ti. A relative density of 97% can be obtained for TiB_2-$(Al,Ni)_3Ti$ composite with 30wt% $(Al,Ni)_3Ti$.

Figure 5 shows the dependence on Al_3Ti or $(Al,Ni)_3Ti$ content of the Vickers hardness (Hv) of TiB_2-Al_3Ti or TiB_2-$(Al,Ni)_3Ti$ composites sintered at 1273K. The Hv of TiB_2 sintered without Al_3Ti is as low as 20. This implies that sintering proceeds scarcely at 1273K, probably because the sintering temperature is too low in comparison with the melting temperature of TiB_2. The Hv increases rapidly with the addition of Al_3Ti. The Hv of TiB_2-Al_3Ti composites reaches a maximum value of 2100 for 30 wt% Al_3Ti and decreases for larger additions. The Hv value of Al_3Ti sintered at 1273K is 600, which is considerably smaller than that of TiB_2. Thus, increasing the Al_3Ti content beyond 30 wt% leads to a decrease of Hv. High values of the Vickers hardness have been obtained for TiB_2-$(Al,Ni)_3Ti$ composites, even though the values are lower than those of TiB_2-Al_3Ti composite. The difference may be caused by the fact that the hardness of cubic $(Al,Ni)_3Ti$ is lower than that of tetragonal Al_3Ti.

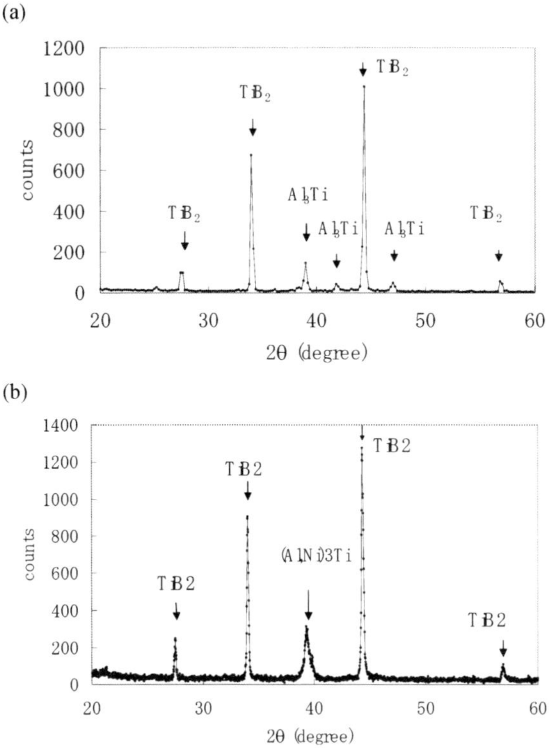

Figure 1. X-ray diffraction patterns of (a) sintered TiB_2:30%Al_3Ti and (b) TiB_2:30%$(Al,Ni)_3Ti$.

Figure 2. SEM image of TiB_2 spark plasma sintered at 1273K.

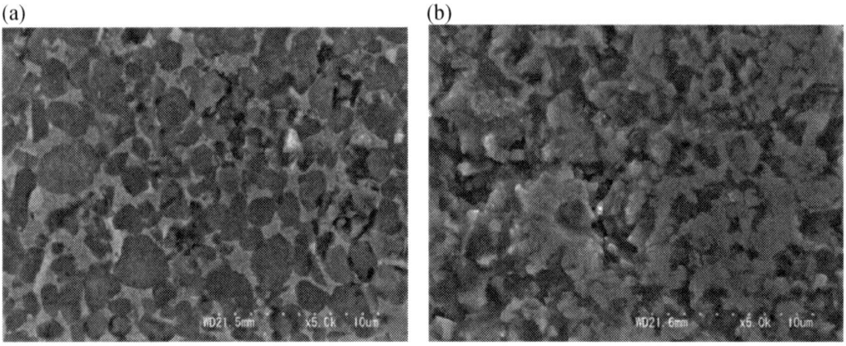

Figure 3. SEM images of TiB_2:30% Al_3Ti (a) and TiB_2:30% $(Al,Ni)_3Ti$ composites fabricated by spark plasma sintering at 1273K.

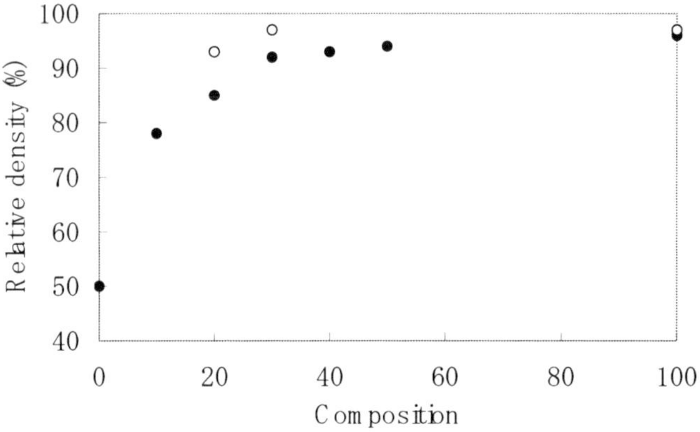

Figure 4. Relative density of the TiB_2-Al_3Ti (●) and TiB_2-$(Al,Ni)_3Ti$ (○) composite produced by spark plasma sintering at 1273K.

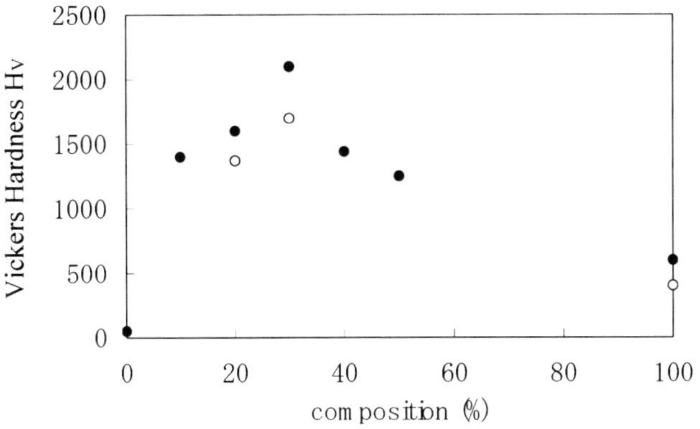

Figure 5. Vickerss hardness of the TiB_2-Al_3Ti (●)and TiB_2-$(Al,Ni)_3Ti$ (○) composite samples sintered by the SPS at 1273K.

In the present study, it has been shown that the hardness, Hv, of TiB_2-Al_3Ti composites sintered at 1273K increases rapidly with the amount of Al_3Ti. Fu et al. have calculated the work of adhesion between TiB_2 and Al_3Ti and have found a good binding between them [2]. The present SEM observations show that Al_3Ti fills in the space between TiB_2 grains (see Figures 3a and 3b), and, therefore, suggest that the Al_3Ti acts as a binder for TiB_2 grains. The Hv of the composite increases as high as 2100 for the composition of 30 wt% Al_3Ti, which is close to the value extrapolated linearly from the Hv value of Al_3Ti (Hv=600) and the reported value for TiB_2 (Hv=3400) [1]. Such high value of hardness of the composite is expected only in the case that direct links between TiB_2 grains are formed. In fact, formation of direct links has been confirmed by the SEM observations presented in Figure 3a. This rapid increase in Hv with the addition of Al_3Ti can be explained by assuming that the diffusion of TiB_2 is accelerated by the presence of Al_3Ti, which causes the fusion of TiB_2 grains and the formation of direct links between TiB_2 grains.

CONCLUSIONS

Sintering of TiB_2 is remarkably improved at 1273K by the addition of Al_3Ti and/or $(Al,Ni)_3Ti$. The microhardness values of TiB_2-Al_3Ti composites with 30 wt% Al_3Ti is as high as 2100 Hv. In contrast sintering of TiB_2 without Al_3Ti additions results in microhardness values as low as 20 Hv. The Al_3Ti has two roles: (1) it fills the space between TiB_2 grains and act as a binder, and (2) it accelerates diffusion of TiB_2 so that strong links between TiB_2 grains are formed.

ACKNOWLEDGMENTS

This work is supported by a Grant-in-Aid for scientific research from, the Ministry of Education Culture, Sports, Science and Technology in Japan.

REFERENCES

1. G.V.Samsonov and I.M.Vinitskii, *Handbook of Refractory Compounds*, IFI/Plenum, New York, 1980.
2. G.Wen, S.B. Li, B.S. Zhang and Z.X. Guo, *Acta Mater.* 49, 1463 (2001).
3. W. Wang, Z.Fu, H.Wang and R.Yuan, *J. Europ. Cer. Soc.* 22, 1045 (2002).
4. M. A.Einarsrud, E.Hagen, G.Pettersen and T.Grande, *J.Am. Ceram. Soc.* 80, 3013 (1997).
5. J.H.Park, Y.H.Lee, Y.H.Koh, H.E.Kim and S.S.Baek, *J. Am. Cer. Soc.* 83, 154 (2000).
6. M. Yamaguchi, H.Inui and K.Ito, *Acta Mater.* 48, 307 (2000).
7. M.Yamaguchi, Y.Umakoshi and T.Yamane, *Philos. Mag.* A55, 301 (1987).
8. Y.Umakoshi, M.Yamaguchi, T.Sakagami and T.Yamane, *J. Mater. Sci.* 24, 1599 (1989).
9. S.M.Kim, M.Kogachi, A.Kameyama and D.G.Morris, *Acta Metall. Mater.* 43, 3139 (1995).
10. Z. Y.Fu, X.M.Min and R.Z.Yuan, *Ceramic Transaction Functionally Gradient Materials* 34, 411 (1992).
11. G. Bernard-Granger and C. Guizard, *Acta Mater.* 55, 3493 (2007).

Mater. Res. Soc. Symp. Proc. Vol. 1373 © 2012 Materials Research Society
DOI: 10.1557/opl.2012.306

Effect of the linear velocity during the melt spinning process on shape memory transformation of Ni-Ti ribbons

E. Nuñez-Mendoza [1-2], E. López Cuellar [1-2], Walman Benicio de Castro[3], B. López Walle [1-2]

[1] CIIDIT, Universidad Autónoma de Nuevo León, parque PIIT km. 10 autopista al Aeropuerto Internacional de Monterrey, Apodaca, N.L., México Cd.
[2] División de Estudios de Posgrado, FIME, Universidad Autónoma de Nuevo León, Ciudad Universitaria, C.P. 66451, San Nicolás de los Garza, N.L., México.
[3] Universidade Federal de Campina Grande, Aprigio Veloso 882-Bodocongó C.P. 10069, Campina Grande, Paraíba, Brasil.

Abstract

The usual process to produce NiTi shape memory alloys is vacuum induction melting (VIM). Currently a new alternative process to produce NiTi shape memory alloys by rapid solidification structures called Melt Spinning has been studied. In this work, results of ribbons with a chemical composition Ti-55.2 Ni (wt %) alloy prepared by this method are presented. The ribbons are prepared at two different linear velocities: 30 m/s and 50 m/s. After that, samples are heat treated at 350 °C during 1 hour. The alloys are characterized by Scanning Electron Microscopy (SEM), Differential Scanning Calorimetry (DSC) and X-Ray Diffraction. According to the cycled DSC test, transformation peaks are associated with the B2→R→B19′ transformation during cooling and B19′→R→B2 during heating, showing transformation in multi-peaks. The martensite B19′ start (M_s) is varying from 35 to 39°C and the martensite finish (M_f) from 15 to 21°C, 42-47°C for austenite B2 start (A_s) and 65-69°C for austenite finish (A_f) approximately. All analyzed ribbons show very similar values of transformation hysteresis temperatures at 50% of transformation of around 28°C. In order to change solidification rate, linear velocity is varied during the melt spinning process. Results indicate that linear velocity affects directly the temperature of transformation. When the linear velocity is increasing, crystallographic Ti-rich precipitates are developed, but dendritic growth disappears, changing the microstructure and decreasing these transformation temperatures. Then changes in linear velocity can dramatically affect shape memory properties, and in this case a velocity of 50 m/s produces a more homogenous alloy.

Keywords: Shape Memory Alloys, Melt Spinning Process, Martensitic Transformation, R-phase transformation

1. Introduction

NiTi shape memory alloys (SMAs) are interesting materials, due to their shape memory effect (SME) and the superelasticity property. Based on these properties, NiTi shape memory alloys have a large range of applications as smart materials in a variety of fields, such as aerospace, biomechanical, biomedical and microelectronic.
Among the known shape memory alloys, this alloy is the most commonly used because of its excellent mechanical properties, corrosion resistance and biocompatibility [2, 3].
Near equiatomic NiTi alloys are well known to undergo a thermo-elastic martensitic transformation, from the B2 austenite parent phase to the B19′ martensite (monoclinic structure with a_o=2.88 Å, b=4.12, c=4.62). And sometimes, the R-phase transformation is known to

precede the B2-B19′ transformation under certain conditions, i.e., thermo-mechanical treatments of near equiatomic NiTi alloys.
This has led to the study of additions of a third or more elements to the binary Ni-Ti or to the use of nonconventional production techniques, in order to modify the martensitic transformation temperatures, hysteresis, and the thermal stability of the alloys, to improve some mechanical properties [6].

The industrial and laboratory scale production of these materials is carried out using different processes, such as air induction melting (AIM), vacuum induction melting (VIM), vacuum consumable and non-consumable arc melting (VAM) and *rapid solidification* (melt spinning). This last one will be described more deeply in the next paragraph [1]. Although the technique is not new, its application to SMAs is relatively recent. Probably the best advantage of this technique is that almost ready-to-use ribbons are directly produced, avoiding the typical thermo-mechanical shaping procedures needed after conventional casting, thus leading to near-net-shape SMAs for certain applications [6].

Recently, rapid solidification *(by melt spinning)* method is reported to induce B2-R transformation in NiTi alloys and this is attributed to an introduction of crystallographic defects such as dislocations by rapid solidification. According to the previous reports, the R-phase transformation is induced at a linear velocity of 21 m/s in 50 Ti-Ni (at. %) alloys [4]. It is known that austenite (B2 cubic structure with a_o=3.02 Å) to R-phase transformation is a first-order displacive transformation, therefore it is agreed that the rhombohedral R-phase has a hexagonal unit cell with a_o=7.38 Å, c=5.32 Å [5].

Based on lattice parameter, crystallographic system and space group of each phase, structures are built in order to know the X-ray diffraction patterns and furthermore identify the corresponding planes.

The purpose of the present study is to investigate the efficacy of the melt spinning process to produce shape memory alloy ribbons and the effect of the linear velocity used during the process. For this, to samples of NiTi are produced by this process using two different linear velocities: 30 m/s and 50 m/s. Results of microstructure, present phases and temperatures of transformation are studied by SEM, X-Ray diffraction and DSC respectively. Results seem to indicate that linear velocity affects directly the temperature, at which the phase transformation occurs. When the linear velocity is increasing crystallographic Ti-rich precipitates are developed, but dendritic growth disappear, changing the microstructure and decreasing these transformation temperatures. Results show that when the linear velocity is changed, shape memory properties are also changed, and in this case with 50 m/s, it can be obtained a more homogenous alloy with a small grain sizes. This is perfect to control the grain size with previous heat treatments to keep a compromise of grain size with heat treatments and mechanical properties.

2. Experimental

Various types of Ti-55.2 Ni (wt %) ribbons are manufactured by Melt Spinning using two linear velocities: 30 m/s and 50 m/s. The alloy is produced in a crucible at 1350°C with a pressure of 200 mbar, the distance between the injection of the crucible nozzle and the copper-wheel rotating is 0.5 mm with an angle of 5°, obtaining ribbons with 30 μm of thickness and 1.7 mm of width. Thereafter, the melt spinning ribbons NiTi are cut in rectangular specimens and heat treated at 350°C during 1 hour. Several characterization techniques are employed to identify the composition, microstructure, phases present, and transformation behavior of the melt spinning ribbons produced. Differential scanning calorimetry (DSC) is used for establishing the martensite ↔ austenite phase transformation temperatures and enthalpies, DSC is carried out at 10°C/min cooling and heating rate and the temperature scanning range is from -10°C to 90°C, using a Diamond DSC of Perkin Elmer. Scanning electron microscopy (SEM) NOVA NanoSEM-200 is employed to investigate the microstructure and the morphology of the ribbons; in addition, by energy dispersive X-ray spectrum (EDX) the average compositions of NiTi ribbons are determined. Crystallographic planes and phase presents of the ribbons are investigated by X-ray diffractions (XRD) using CuKα radiation, (Bruker Advanced X-ray Solutions D8).

3. Results and discussion

3.1. DSC study of NiTi ribbons produced at different linear velocities

Fig. 1a and b show DSC curves of Ti-55.2 Ni (wt %) alloy ribbons fabricated at two different linear velocities and heat-treated at 350°C during 1 hour. Clear exothermic and endothermic multi-peaks are seen on cooling and heating curves. The formation of these multi-peaks during cooling seems to be caused by a transformation in three steps, from the austenite to R-phase and finally to martensite (B2→R→B19′) and during heating in a reversible process (B19′→R→B2). In table 1 are summarized all the transformation temperature values obtained from DSC curves.

As can be seen, the transformation temperatures are different when the linear velocity changes; for example transformation temperatures of B19′and B2 phases are greater in those samples produced at 30 m/s. In the ribbons produced at 30 m/s the temperatures A_s (the temperature at which transformation to the austenite phase starts) is 58°C and A_f (the temperature at which complete transformation to the austenite phase is reached), is determined to be 69°C with an enthalpy of transformation (about 24.7 J/g). During the cooling multi-peaks take place, first the R-phase appear with start and finish transformation temperature values, being R_s=42°C and R_f=31°C, for the martensitic phase M_s=31°C and M_f=21°C (M_s=martensite phase starts and M_f=martensite phase finish), but the enthalpy necessary to achieve the transformation is reduced at 18.1 J/g approximately.

If the DSC curves corresponding to the ribbons fabricated at 50 m/s are observed, a decrement in the transformation temperatures A_s and A_f, being A_s=54°C and A_f=65°C, besides the R-phase also is present during the cooling, but now R_s=40°C and R_f=27.5°C, In other hand, the temperatures of martensite transformation also decreasing various units in this case M_s=27°C and M_f=15°C, the enthalpies of transformation for this ribbons are 23.6 J/g for B2 phase and in the case of martensite B19′ the enthalpy is reduced due to the presence of the R-phase. The temperature values of the R-phase during the heating does not presented because the peaks are not enough clear in comparison with the cooling scanning, this for both linear velocities at which the ribbons are produced. These changes in temperature are attributed to the fact that solidification rate of the wheel over the ribbons in the melt spinning process is higher, beside, is known that crystallographic defects have an important influence in the thermal behavior of shape memory alloys.

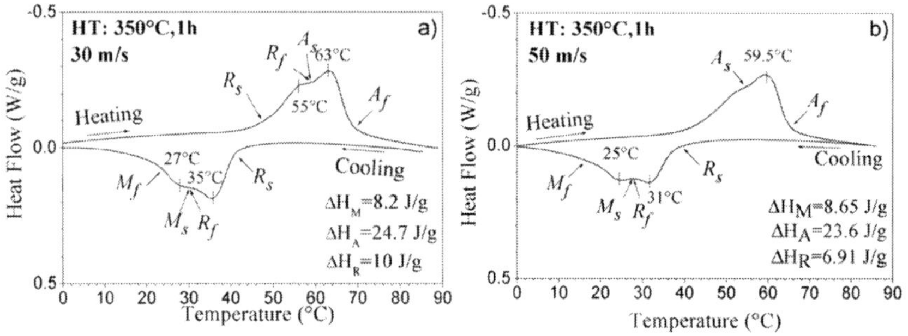

Figure 1. DSC curves of NiTi alloy ribbons fabricated at different linear velocities, a) 30 m/s and b) 50 m/s, both heat-treated at 350°C × 1h.

Table 1
DSC results (°C)

Ribbon	A_s	A_f	M_s	M_f	R_s	R_f
HT:350°C,1hr,30m/s	58	69	31	21	42	31
HT:350°C,1hr,50m/s	54	65	27	15	40	27.5

3.2. X-ray diffractions patterns of NiTi ribbons

In order to explain DSC curves of Fig. 1, X-ray diffractions are made at room temperature, and the results obtained are shown in Fig. 2.

Figs. 2a and b show X-ray patterns of the Ti-55.2 Ni (wt %) alloy ribbons fabricated at 30 m/s and 50 m/s respectively. In the X-ray patterns at 30 m/s, diffraction peaks corresponding to the B19′ phase only are found. This is normal because at room temperature, the alloy has almost been all transformed in to martensite, as can be observed in DSC curves. Contrary to results of ribbon fabricated at 50 m/s observed in figure 2b, where B19′ and R-phase peaks are detected, which also is expected because if DSC results are observed, at room temperature not all the alloy has been transformed into martensite, i.e. an important amount of the material still present in R-pase. So, X-ray results confirm that the transformation observed in multi-peaks in DSC curves can be attributed to the presence of the R-phase. The crystal planes are indexed based on works developed by Tae-Hyun Nam and co-workers (2002), Ken K. Ho and Gregory P. Carman (2000), beside, based on lattice parameters and space groups of each phase are built crystallographic structures by CaRIne software with the purpose of identify the crystal planes.

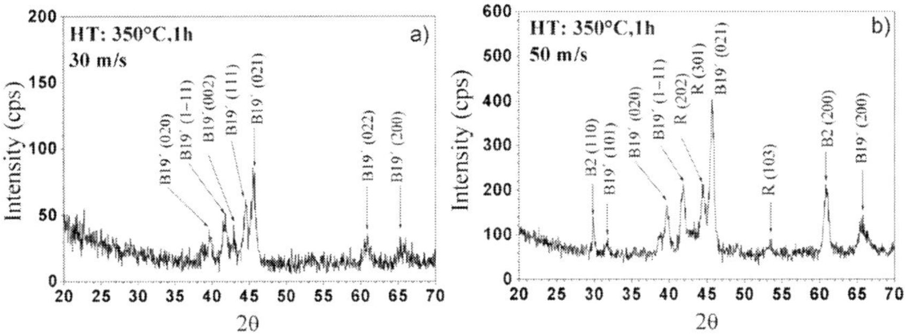

Figure 2. X-ray patterns of Ti-55.2 Ni (wt %) ribbons a) produced at 30 m/s and b) 50 m/s; both are heat-treatment at 350°C × 1h.

3.3. Microstructure and compositional analysis by SEM

Fig. 3a corresponding to NiTi alloy ribbons produced at 30 m/s. Image shows dendritic growth over the grain boundaries with grains between 10-30 μm. So dendritic growth promotes changes in compositions and in properties, which leads to a heterogeneous alloy, and different grain size also leads to a heterogeneous microstructure. In contrast if the Fig. 3b is observed, which corresponds to the ribbon produced at 50 m/s, it can be clearly seen that the microstructure presents a more homogenous grain size of around 800 nm with some Ti-rich precipitates, identified as by EDX (Energy Dispersive X-ray). In this sense, when the linear velocity increases, also the cooling rate is increased, and this gives less time to the formation of dendrites and to the grain growth, promoting in this case a more homogenous alloy. In other hand, the decrease of the transformation temperatures of the sample at 50 m/s, can be attributed to the smaller grain size and the Ti-rich precipitates found in its microstructure [7].

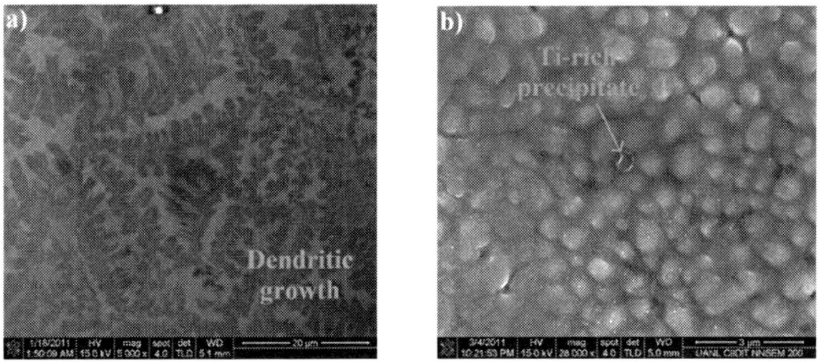

Figure 3. SEM micrograph ribbons a) produced at 30 m/s showing dendritic growth and b) 50 m/s, which present dendritic growth and Ti-rich precipitates.

4. Conclusions

Ribbons of NiTi produced by melt spinning process present the shape memory effect, independently of the linear velocity at which are made, validating that this process is a good alternative to produce ribbons with the shape memory effect. During the transformation from austenite to martensite, the R-phase has been detected. The linear velocity of the melt spinning process has an important effect over the thermal behavior of the ribbons, in this case increasing the linear velocity from 30 to 50 m/s, transformation temperatures have a tended to decrease, due the presence of crystallographic defects such as smaller grain size and Ti-rich precipitates. Moreover, when the linear velocity increases, also the cooling rate is increased, and this gives less time to the formation of dendrites and to the grain growth, promoting in this case a more homogenous alloy. Finally, after these results the sample produced at 50 m/s is more interesting due to its homogeneity and their small grain size, because it will be easy to control the grain size, simply with heat treatments, and to establish a compromise between grain size, heat treatments, and mechanical properties.

Acknowledgements

Authors wish to thank to the CONACYT and CIAM for their financial support from the 82515, 81546 and 104310 projects. This work is technical supported by FIME-Universidad Autónoma de Nuevo León (UANL), Centro de Innovación, Investigación y Desarrollo en Ingeniería y Tecnología (CIIDIT) and Department of Mechanical Engineering, Universidade Federal de Campina Grande, Brazil.

References

1. C. J. de Araújo, A. A. C. Gomes, J. A. Silva, A. J. T. Cavalcanti, R. P. B. Reis, C. H. Gonzalez, Journal of Materials Processing Technology, Vol. 209, pp. 3567-3664 (2009).
2. D. Batalu, H. Guoqiu, a. Aloman, L. Xiaoshan, Z. Zhihua, Journal of Optoelectronics and Advanced Materials, Vol. 8, pp. 694-698 (2006).
3. W. C. Crone, D. Wu, J. H. Perepezko, Materials Science and Engineering, Vol. A 375-377 pp. 1177-1181, USA (2004).
4. Tae-Hyun Nam, Jae-Hoon Kim, Mi-Seon Choi, Journal of Materials Science Letters, Vol. 21, pp. 685-688 (2002).
5. Q. Chen, K. M. Knowles, C. J. Humphreys, X. F. Wu, Journal of Materials Science, Vol. 31, pp. 4227-4231 (1996).
6. R. Santamarta, E. Cesari, J. Pons, and T. Goryczka, Metallurgical and Materials Transactions, Vol. 35A, pp. 761-770 (2004).
7. K. N. Lin, S. K. Wu, Journal of Alloys and Compounds, Vol. 424, pp. 171-175 (2006).

Mater. Res. Soc. Symp. Proc. Vol. 1373 © 2012 Materials Research Society
DOI: 10.1557/opl.2012.307

The effect of rapid solidification and grain size on the transformation temperatures of Ni-44,8wt%Ti melt spun alloy

G. C. S. Anselmo, W. B. de Castro*; C. J. de Araújo
UAEM, Universidade Federal de Campina Grande, Aprígio Veloso, 882 – Caixa *Posta 10069 - 58429-900 – Campina Grande - PB – Brazil
georgeanselmo@yahoo.com.br, walman@dem.ufcg.edu.br*, carlos@dem.ufcg.edu.br

ABSTRACT

Ribbons of the Ni-44.8wt%Ti shape memory alloy are prepared through the melt spinning technique. The study is focused on investigating the effect of the rapid solidification and grain size at characteristic start martensitic (M_s), final martensitic (M_f), start austenite (A_s) and final austenite (A_f) transformation temperatures. Changes on martensitic transformation temperatures in $Ti_{45}Ni_{55}$ melt spun ribbons are observed as grain size is reduced. Results of optical microscopy and differential scanning calorimetry (DSC) are used to associate grain size with transformation temperatures.

Keywords: Shape memory Ni-Ti alloy; Rapid solidification; Martensitic transformation.

INTRODUCTION

Shape memory alloys (SMAs) represent a unique class of materials that undergo a reversible phase transformation (martensitic transformation) allowing these materials to display dramatic pseudoelastic stress-induced deformations and recoverable temperature-induced shape memory deformations. These materials are used as smart materials in a variety of aerospace, biomechanical, and microelectronics applications. Among the known shape memory alloys, Ni-Ti is the most commonly used because of its excellent mechanical properties, corrosion resistance and biocompatibility [1]. In the recent past, the field of microsystems has been subject to growing attention from both the industry and the research community. Microsystems have been recognized as having the potential to revolutionize the performance of a wide range of products by merging silicon-based microelectronics with micromachining technologies, thus enabling complete systems-on-a-chip to be developed and allowing novel functionalities at reduced costs. In this context, the application of shape memory alloys for actuation of micropneumatic devices might bring a relevant technological breakthrough. SMA materials exhibit the highest energy density amongst current micro-electromechanical systems MEMS compatible materials and, more importantly, as size is reduced towards the micro-scale, they benefit from improved heat transport, which increases their response speed [2].

Many studies have been undertaken to find a method to control the martensitic transformation. According to previous studies, it is significantly affected by the alloy composition, crystallographic defects such as dislocations, precipitation and grain size [3]. Grain boundaries are believed to strengthen parent phases, and therefore Ms decreases with decreasing grain size. In Ti–Ni-based alloys, the critical austenite grain size for martensitic transformation is known to be 50 nm, below which martensitic transformation does not occur.

The aim of this work is to study the relationship between grain size and martensitic transformation temperature in the Ni-44,8wt%Ti alloy. In order to attain a wide range of cooling rate and grain size, the melt spinning technique is therefore considered a suitable preparation route for this alloy. Melting spinning is an important method for producing metals with improved mechanical and/or physical properties. This technique employs a very high cooling rate (up to 10^6 °C/s). In general, such a high cooling rate has the advantage of refinement of grain sizes.

EXPERIMENTAL

The Ni-44,8wt%Ti ingot with 19 mm in diameter by 100 mm long is prepared using the conventional vacuum arc-remelting (VAR) method. High purity Ti and Ni raw materials are repeatedly melted six times in an argon atmosphere for homogenization. Then they are cut into small pieces, each of which has a weight in the range of 10 to 30 grams, induction-melted in an argon atmosphere in a quartz crucible at 1250 °C supplied into a single-roller melt-spinning machine and subsequently ejected by high pressurized argon out of a 0.4 mm orifice onto a 200 mm diameter copper roller with at different tangential speeds between 30 and 50 m/s. The final ribbons of the melt-spinning process ranged between 30 μm and 41 μm in thickness and 1 mm in width. Transformation temperatures and enthalpies of as-spun ribbons are determined by differential scanning calorimetry TA Q20 DSC with 10 °C/min heating and cooling rate and the temperature scanning range is from −30 to +200 °C. The software for DSC for data acquisition, storage and evaluation under MS WINDOWS. It has multitasking with simultaneous operation of several thermal analysis systems and simultaneous evaluation graphical user surface, integrated detailed HELP system. Microstructure of cross-section of ribbons is examined by Nikon FX-35DX optical microscope (OM). For optical microscopy, the specimens are polished using conventional procedures and etched in a solution composed of $HF:HNO_3:H_2O$= 4:5:10 (in volume). The etching time is about 10–15 s. From the OM images under 1000× magnification, the average grain size of ribbons is estimated by the linear intercept method. The numbers of intercepted grains are at least 10 and 50 for large and small grains, respectively.

RESULTS AND DISCUSSION

The increase of the wheel velocity from 30 to 50 m/s results in a decrease of the ribbon thickness from 41 μm to 30 μm. As the increase of the wheel velocity leads to a reduced ribbon thickness, the heat transfer coefficient at the quenching wheel–ribbon interface is enhanced and the cooling rate increases. Therefore, this result clearly indicates that variations in the melt spinning temperature and wheel velocity allow the effective control of the cooling rates in the melt spinning process.

Figure 1 shows optical micrographs of as-spun ribbons fabricated at two different wheel velocities of 30 and 50 m/s. As seen in Fig. 1, as-spun Ni-44.8wt%Ti ribbons are fully crystallized and most of the columnar and small grains are located at the free surface and copper roller surface of the ribbons, respectively. Refined grain structure is the hypothesis to explain the decrease in the transformation temperature, as shown in Fig. 1. When the ribbon is produced at a higher wheel velocity in melt spinning, the degree of undercooling becomes high because of its thinner thickness and the amount of crystalline layer decreases with wheel velocity. Variations in grain size are often accompanied by changes in dislocation structure and precipitation, which also greatly affect the transformation temperatures. Therefore, it seems to be difficult to investigate the effect of grain size on the transformation temperatures in Ti–Ni alloys. Grain refinement has been reported in literature to decrease the transformation temperatures [4].

(a)

(b)

Fig. 1. Optical micrographs of the ribbons (light brown) fabricated at the wheel velocities of (a) 30 m/s and (b) 50 m/s.

Figure 2 shows the DSC curve of as-spun grain-size mixed Ni-44.8wt%Ti ribbons. As can be seen, the transformation peaks are short with large transformation enthalpies for ribbons that are fabricated with wheel velocities of 30 m/s. Broad transformation peaks with short transformation enthalpies correspond to ribbons that are fabricated with wheel velocities of 50 m/s. This is most likely due to a greater amount of defects and residual stress for wheel velocities of 50 m/s. At the same time, the grain size inherent in ribbons is finer when higher speeds are used, as shown in Fig. 1. The grain boundaries and defects can act as barriers to the martensitic transformation as a result of the extra energy required during transformation [5]. Thus fine-grain ribbons with a higher amount of grain boundaries are expected to have lower transformation temperatures and smaller transformation enthalpies, as shown in Fig. 2.

NiTi shape memory alloys (SMA) transform martensitically from a B2 cubic austenite into a monoclinic B19' martensite either directly or via a rhombohedral R-phase martensite. The B19' martensite can be obtained either by a single step transformation of B2 → B19', or by a two-step transformation of B2 → R-phase → B19' [8]. Note in Fig. 2 that when ribbons are fabricated with wheel velocities of 30 m/s there are one B2 ⇨ B19' martensitic transformation peak in cooling and heating, but when ribbons are fabricated with wheel velocities of 50 m/s there are B2 ⇨ R and R ⇨ B19' martensitic transformation peaks in cooling and heating and B19' ⇨ R and R ⇨ B2. The occurrence of the multiple martensitic transformations in ribbons of Ti51Ni49 SMA is due to the coexistence of large and small grains distributed in the ribbons. The reason why both R→B19' and B19'→B2 transformations are separated into two peaks which correspond to large and small grains, while the B2↔R transformation is not separated into two peaks is that the B2↔R transformation exhibits much smaller transformation strain than R→B19' and B19'→B2 transformations.

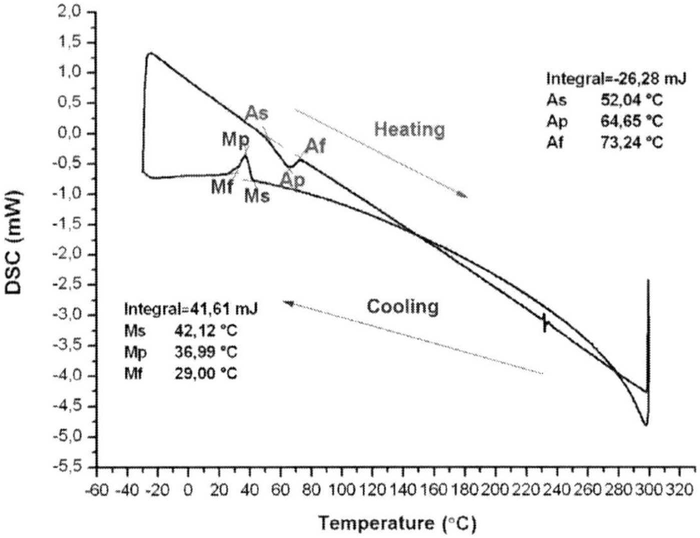

(a)

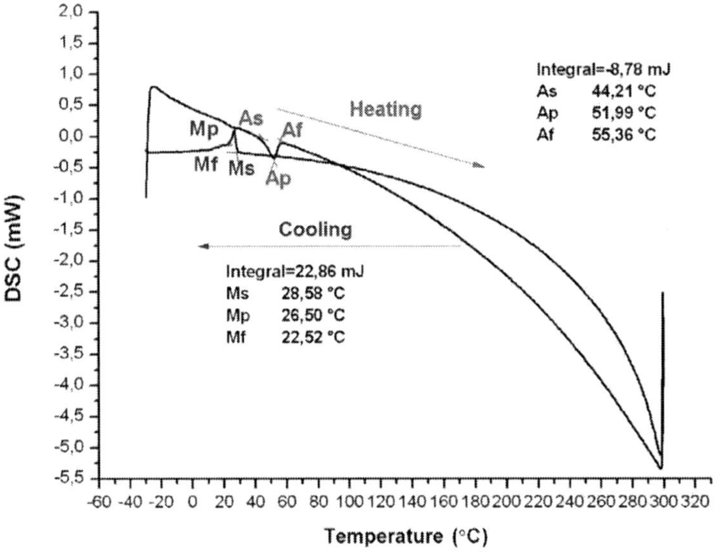

(b)

Fig. 2 - DSC curves of the ribbons fabricated at different wheel velocities: (a) 30 m/s and (b) 50 m/s.

Fig. 2 shows a fully crystallized and most of the columnar and small grains are located at the free surface and copper roller surface of the ribbons, respectively. Then, the multi-stage martensitic transformation induced by the inhomogeneous grain size distribution in the ribbon is observed in this study [1]. According to the reported studies, the R-phase appearing in the martensitic transformation of Ti-rich TiNi thin films is induced either by the coherent stress field around the plate-like Guinier-Preston (GP) zones or by the semicoherent stress field around the spherical Ti_2Ni precipitates [6-9]. Increasing wheel velocities caused the formation of the R-phase. The as-spun $Ti_{45}Ni_{55}$ ribbon has many GP zones, since rapid solidification produces an increase in the number of defects [10] inducing the formation of R-phase.

CONCLUSIONS

Transformation behavior of melt spun Ni-44,8wt%Ti alloy ribbons fabricated at different cooling rates by melt spinning is investigated. The increase of the wheel velocity from 30 to 50 m/s results in a decrease of the ribbon thickness. When the ribbon is produced at a higher wheel velocity in melt spinning, the degree of undercooling becomes high because of its thinner thickness. The grain boundaries and defects can act as barriers to the martensitic transformation as a result of the extra energy required during transformation. Thus fine-grain ribbons which have lots of grain boundaries would be expected to have lower transformation temperatures. The DSC result showed that when wheel velocity increases from 30 m/s to 50 m/s the R-phase appears because the as-spun Ni-44,8wt%Ti ribbon has many Guinier-Preston (GP) zones, induced by the large number of defects. According to the partial-cycled DSC test, we conclude that the peaks of transformations are associated with B2→R transformation, R → B19' transformation for large grains, R → B19' transformation for small grains during cooling, and B19' → B2 transformation for large grains and B19' → B2 transformation for small grains during heating. The occurrence of the multiple martensitic transformations in ribbons of $Ti_{45}Ni_{55}$ SMA is due to the coexistence of large and small grains distributed in the ribbons. The reason why both R→B19' and B19'→B2 transformations are separated into two peaks which correspond to large and small grains, while the B2↔R transformation is not separated into two peaks is that the B2→R transformation exhibits much smaller transformation strain than R→B19' and B19'→B2 transformations.

ACKNOWLEDGEMENTS

The authors would like to acknowledge the financial support from the CNPq, from CASADINHO project nº 620091/2008-8, from Projeto Universal project nº 471831/2009-3 and the concession of scholarship to George Carlos S. Anselmo.

REFERENCES

[1] Y. Freed, J. Aboudi, International Journal of Solids and Structures, vol. 46, 2009, pp. 1634–1647.
[2] Y. Bellouard, Materials Science and Engineering, vol. A 481–482, 2008, pp. 582–589.
[3] H. Morawiec, J. Lelatko, D. Stróz, M. Gigla, Materials Science and Engineering, vol. A273–275, 1999, pp. 708–712.
[4] K. N. Lin, and S. K.Wu, Journal of Alloys and Compounds, vol. 424, 2006, pp.171–175.
[5] L. Zhang, C. Xie, J. Wu, Journal of Alloys and Compounds, vol. 432, 2007, pp. 318–322.

[6] X. Zhang and H. Sehitoglu, Materials Science and Engineering, vol A 374, 2004, pp. 292–302.
[7] P. Sittner, M. Landa, P. Lukás and V. Novák, Mechanics of Materials, vol. 38, 2000, pp. 475–492.
[8] Y. Zhou, J. Zhang, G. Fan, X. Ding, J. Sun, X. Ren, K. Otsuka, Acta Materialia, vol. 53, 2005, pp. 5365–5377.
[9] R. Zarnetta, D. Kõnig, C. Zamponi, A. Aghajani, J. Frenzel, G. Eggeler, A. Ludwig, Acta Materialia, vol 57, 2009, pp. 4169–4177.
[10] L. Wu, S. Chang and S. Wu, Journal of Alloys and Compounds, vol. 505, 2010, pp.76–80.

Mater. Res. Soc. Symp. Proc. Vol. 1373 © 2012 Materials Research Society
DOI: 10.1557/opl.2012.308

Rotating Bending Fatigue Endurance and Effect of Controlled Corrosion on the AISI-SAE 6063-T5 Aluminum Alloy.

G. M. Domínguez Almaraz[1], J. L. Ávila Ambriz[1], E. Cadenas Calderón[1], J. J. Villalón López[1]
[1]Facultad de Ingeniería Mecánica, Universidad Michoacana de San Nicolás de Hidalgo (UMSNH), Santiago Tapia No. 403, Col. Centro, Morelia, Michoacán 58000, México.

ABSTRACT

Rotating bending fatigue test are carried out on the aluminum alloy 6063-T5 for corroded and non corroded specimens. Special attention is devoted to fatigue endurance reduction caused by controlled surface corrosion on corroded specimens. Corrosion attack is implemented by submersion of specimens in an acid solution for: two, four and six minutes in order to induce three degrees of surface corrosion. The corrosion agent is a solution of hydrochloric acid with a PH close to 0.8 and solution concentration of 38%. Rotating bending fatigue tests at frequency of 50 Hz, room temperature and without environmental humidity control are carried out on 4 types of specimens: without corrosion and 2, 4, and 6 minutes immersed in the solution of hydrochloric acid. Results are analyzed regarding the corrosion effect on fatigue endurance and conclusion are enlisted concerning rotating bending fatigue tests and corrosion attack on this aluminum alloy.

INTRODUCTION

Industrial applications of aluminum alloy 6063-T5 are mainly oriented to: aluminum extruded parts in car industry, architectural extrusions, irrigation system pipelines, electrical conductors [1,2,3]. This alloy presents high surface polish, good welding properties, corrosion resistance and it is typically produced with very smooth surfaces fit for anodizing. Usually, this aluminium alloy undergoes mechanical and environmental loads (corrosion) in most of its applications [4] and some coating development has been introduced recently in order to improve its mechanical and physical properties [5]; thus, the understanding behavior of this aluminum alloy undergoing mechanical loading and corrosion attack is of principal interests. In the present study rotating bending fatigue tests are carried out on the hourglass shape specimens at the frequency of 50 Hz, without control of environmental humidity and surface roughness; nevertheless, all specimens are machined similarly in order to maintain the surface roughness with no large variation.

EXPERIMENTAL DETAILS

The rotating bending fatigue machine is shown in Figure 1. It has been developed in our laboratory in order to obtain the fatigue endurance on this aluminium alloy [6]. The high controlled speed of this machine is 150 Hz and stops automatically when the specimen is broken. Additionally, an incorporated electronic system is designed to count the number of cycles and stops simultaneously when the specimen is broken.

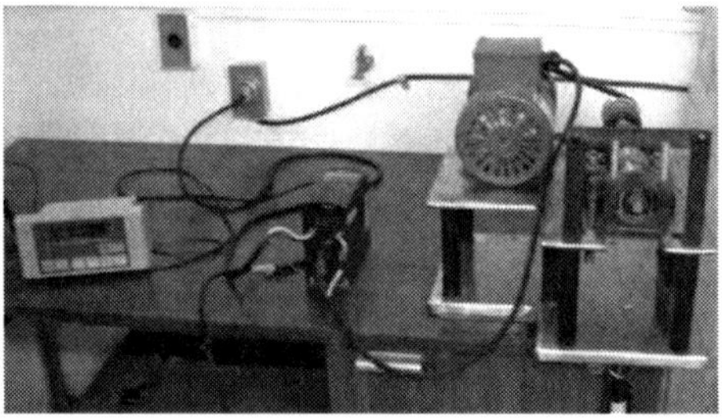

Figure 1. Patented rotating bending fatigue machine.

Chemical composition in weigh and principal mechanical properties is shown in Tables 1 a) and b) respectively.

Table 1. a) Chemical composition of aluminum alloy 6063-T5.

Si	Fe	Cu	Mn	Mg	Cr	Zn	Tl	Al
0.3-0.6	0.35	0.1	0.1	0.4-0.85	0.1	0.1	0.1	> 96.9

Table 1. b) Principal mechanical properties of aluminum alloy 6063-T5.

Elastic Limit MPa	UTS Mpa	% elongation	E Young Gpa	Hardness Brinell
145	187	0.33	68.9	60

Experimental results are plotted in Figure 2. Corrosion attack induces an important decrease of fatigue endurance of this aluminum alloy; the non corroded specimens show higher fatigue lives as compared with the corroded specimens for the 4 values of applied load σ_a: 60%, 70%, 80% and 90% regarding the elastic limit of this material.

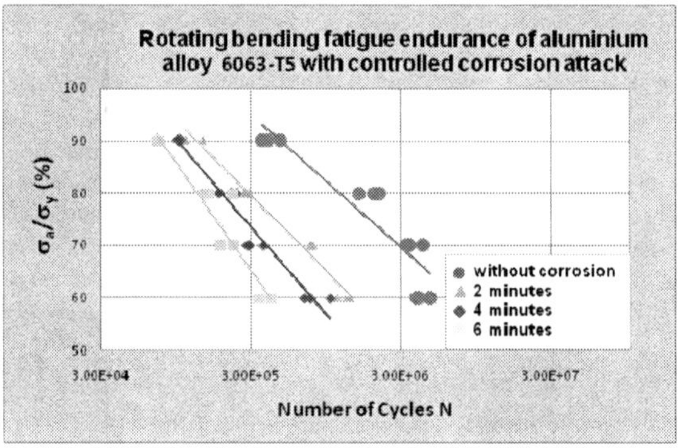

Figure 2. Fatigue life and tendency curves for 4 types of aluminum specimens, under rotating bending fatigue tests.

Concerning the corroded specimens, a long time of attack corrosion is translated into a decrease of fatigue life and this difference is higher for the low loading regimes (60% and 70% of elastic limit).

DISCUSSION

Fatigue endurance on metallic elements undergoing mechanical loading is reduced notably with corrosion attack [7, 8, 9]. For a given loading ratio σ_n/σ_y, see Figure 2, the fatigue endurance is reduced for the highest time of corrosion attack on testing specimens. The presence of corrosion pitting holes at the specimen surface, Figure 3, is related to stress concentrations factors accelerating the fracture initiation and crack propagation [10, 11]. In addition to the size of the corrosion pitting holes, the proximity of two or more of them increases the stress concentration factor and accelerates the crack initiation and propagation [12]. Concerning the average size of pitting holes on corroded specimens, it is comprised approximately between 150 μm and 250 μm for the 2 and 4 minutes of corrosion. Pitting holes density on specimen surface increases with time of corrosion attack rather than increasing the pitting holes size. This tendency is observed on Figure 3 c); here, the longest time of corrosion attack induces coalescence of contiguous pitting holes forming "grooves" along the longitudinal direction of specimen.

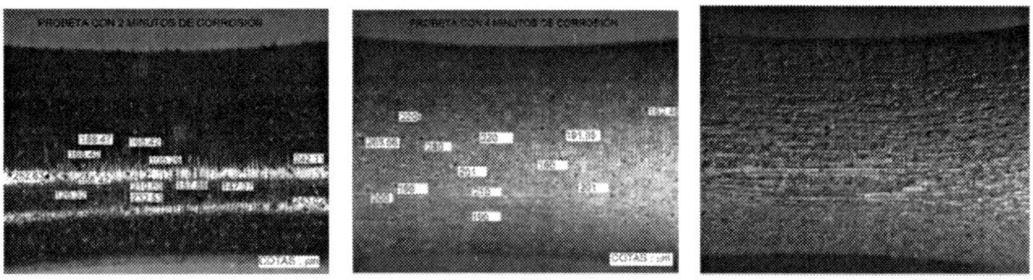

a) b) c)

Figure 3. Corroded specimens before crack with: 2 minutes a), 4 minutes b), and 6 minutes of corrosion c).

Reduction of fatigue endurance is observed to be associated with pitting holes. Crack initiation is identified with the presence of one, two or more neighboring corrosion pitting holes. With the analysis of fracture surfaces and the related pitting holes, a modified Kitagawa-Takahashi diagram is constructed taking one or more pitting holes at crack initiation as the crack length "a", see Figure 4. The non corroded specimens provide information for the very small crack sizes; for 60% of elastic limit loading: σ_a = 87 MPa, $\Delta\sigma_{fat}$ = 174 MPa with R=-1, the fatigue life is N ≈ 4 x 10^6 cycles, see Figure 2. The threshold stress-intensity range for fatigue-crack growth is ΔK_{TH}, and it is defined as the stress-intensity range below which cracks appear do not develop, operationally defined here when da/dN < 10^{-7} mm/cycle. Stress-intensity range at threshold is obtained for aluminum alloy 6063-T5 with R = 0.1 [13], the value is ΔK_{TH} = 3 MPa $m^{-1/2}$; this value should be close to 4 MPa $m^{-1/2}$ for R = -1.

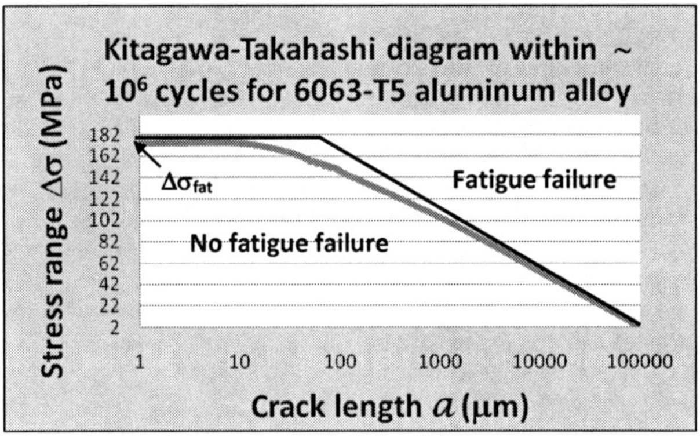

Figure 4. Kitagawa-Takahashi diagram for aluminum alloy 6063-T5.

ΔK_{TH} does not depend on crack size and at small crack sizes this parameter tends to a limiting condition given by the fatigue endurance strength, which for this aluminum alloy is $\Delta\sigma_{fat}$ = 174 MPa within N ≈ 4 x 10^6 cycles. Taking a_0 as the crack length delimiting the zone which the fatigue threshold is no longer crack size independent; then, the variation of stress range may be evaluated as a function of additional crack length "a" using the El Haddad et al.'s single empirical equation [14]:

$$\Delta\sigma = \frac{\Delta K_{TH}}{Y\sqrt{\pi(a + a_0)}} \qquad (1)$$

Here, Y is the geometry factor that for these specimens can be taken as: Y = 0.97. From eq. (1), it is possible to evaluate a_0 by:

$$a_0 = \left(\frac{\Delta K_{TH}}{Y\Delta\sigma_{fat}}\right)^2 \frac{1}{\pi} \qquad (2)$$

After substituting the values for this aluminum alloy, a_0 ≈ 179 μm, this point is identified in the decreasing red line in Figure 4.

CONCLUSIONS

- Rotating bending fatigue endurance decreases with the time of corrosion attack, in corroded specimens.
- The reduction on fatigue endurance increases when the loading rate decreases (for 60 and 70% of elastic limit of material).
- Corrosion pits size stay approximately unchanged with corrosion time, corrosion pits density in surface specimen increase with corrosion time.
- A Kitagawa-Takahashi diagram is constructed for this aluminum alloy in the fatigue life close to N ≈ 4 x 10^6 cycles. Under this condition, stress range for non corroded specimens is found to be: $\Delta\sigma_{fat}$ = 174 MPa for R = -1.

- Analysis on fracture surfaces shows that corrosion pitting holes are the sites for crack initiation and propagation. Crack initiates at pits and propagates following the path of close pits, as shown on Figure 5 a), b) and c). Sometimes, parallel secondary propagation paths are observed on fracture surface; theses ones competing with principal fractures paths during testing.
- Frontal fracture surface, Figure 5 d), show the typical fracture under rotating bending fatigue: high plastic deformation close to specimen surface where high stresses are acting (crack initiation zone) and granular area at the center characterizing the fast crack propagation. Often, more than one crack initiation point is identified at fracture surface.

a) 6 min. of corrosion, 70% of elastic limit

b) 4 min. of corrosion, 90% of elastic limit

c) 2 min. of corrosion, 90% of elastic limit

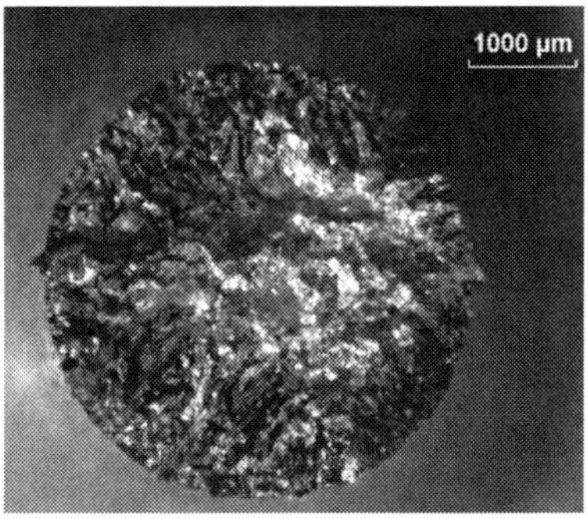

d) 4 min. of corrosion, 70% of elastic limit

Figure 5. Images of fracture surfaces.

ACKNOWLEDGMENTS

We express our gratitude to the University of Michoacan (UMSNH) in Mexico for the facilities received to carry out this study. A special mention of gratitude to CONACYT

(The National Counsel for Science and Technology in Mexico) for the financial support destined to this research work.

REFERENCES

1. A.S. Fouda, A.A. Al-Sarawy, F.Sh. Ahmed, and H.M. El-Abbasy, *Corrosion Science*, Vol. 51 (2009), pp. 485-492.
2. J. Hirsch, "Virtual Fabrication of Aluminum Products (Microstructural Modeling in Industrial Aluminium Fabrication Process)", Ed. Wiley-VCH, First Edition (2006), 381 pages.
3. W.S. Miller, L. Zhuang, J. Bottema, A.J. Wittebrood, P. De Smet, A. Haszler, A. Vieregge, *Materials Science & Engineering A280* (2000), pp. 37–49.
4. G.M. Scamans, N. Birbilis, R.G. Buchheit, Shreir's Corrosion, Vol. 3 (2009), pp. 1974-2010.
5. C.J. Villalobos-Gutiérrez, G.E. Gedler-Chacón, J.G. La Barbera-Sosa, A. Piñeiro, M.H. Staia, J. Lesage, D. Chicot, G. Mesmacque and E.S. Puchi-Cabrera, *Surface and Coating Technology*, Vol. 202 (2008), pp. 4572-4577.
6. G. M. Dominguez Almaraz, M. Guzmán Tapia, Patent No. 276818, Mexico (2010); "High Speed Rotating Bending Fatigue Machine".
7. Q.Y. Wang, N. Kawagoishi , Q. Chen,", *Scripta Materialia*, Vol. 49 (2003), 711-716.
8. N. P. Isekar, A. Jyothirmayi, G. Sundararajan, *Intern. Journal of Fatigue*, Vol. 33 (2011), pp. 1268-1276.
9. S. Dey, A. B. Mallick, I. Chattoraj, *Advanced Materials Research*, Vol. 117 (2010), pp. 43-48.
10. J. Słowik, T. Łagoda, *Intern. Journal of Fatigue*, Vol. 33 (2011), pp. 1304-1312.
11. M. Sakane, S. Zhang, T.-J. Kim, *Intern. Journal of Fatigue*, Vol. 33 (2011), pp. 959-968.
12. G.M. Domínguez Almaraz, V.H. Mercado Lemus, J.J. Villalón López, *Procedia Engineering*, Vol. 33 (2011), pp. 805-813.
13. T. Hung Tra, M. Seino, M. Sakaguchi, M Okasaki, *J. of Solid Mechanics and Material Egineering*, Vol. 4 (2010), pp. 840-848.
14. M.H. El Haddad, T.H. Topper and K.N. Smith, Vol. 11 (1979), pp. 573-584.

Mater. Res. Soc. Symp. Proc. Vol. 1373 © 2012 Materials Research Society
DOI: 10.1557/opl.2012.309

Rheology behavior of lubricants and its effect on the quality surface of aluminum and copper cold-drawn wires

R. Cortes Rodríguez [a-b], R. Garza García[c], J. Pérez Patiño[b], E. Reyes Melo[a-b], E. López Cuellar[a-b]

[a] CIIDIT, Universidad Autónoma de Nuevo León, parque PIIT km. 10 autopista al Aeropuerto Internacional de Monterrey, Apodaca, N.L., México Cd.
[b] División de Estudios de Posgrado, FIME, Universidad Autónoma de Nuevo León, Ciudad Universitaria, C.P. 66451, San Nicolás de los Garza, N.L., México.
[c] Magnekon S.A. de C. V., Ave. Rómulo Garza 290, San Nicolás de los Garza N.L. 66493, México

Abstract

The roll of lubricants in the cold-drawn process is very important to obtain a good quality on the surface of aluminum and copper wires. The viscosity of a lubricant is closely related to its ability to reduce friction. When the viscosity of a lubricant is too low, the lubricated component will have inadequate protection and will therefore be subject to excessive wear. When the viscosity of the lubricant is too high, the lubricated component will expend additional energy to complete its task. In this work, the rheology behavior of traditional lubricants for the cold drawn of Al and Cu is determined from experimental data of viscosity Vs shear rate. To evaluate the efficiency of each lubricant, the roughness surface of each wire is measured by Atomic Force Microscopy (AFM). In this way a minimum of roughness in wires corresponds to the viscosity required for each cold-drawn process. It is known that different lubricants are used for the cold drawn of Al and Cu. In this work, a new lubricant developed with the aim to be used in both process is characterized by FTIR, rheometer analysis and AFM. Results have indicated that, this new lubricant with a low viscosity that promotes a lower energy process, also decreases the roughness of Al and Cu wires compared with conventional lubricants, i.e.it has an important influence in the quality of the wires surface. This means that this new lubricant could be used during the process of both metals without making important changes, which means low operations costs and flexibility for the manufacturing plant.

Keywords: cold-drawn process, aluminum, copper, AFM, lubricant.

1. Introduction

Cold-drawn is a metalworking process used to reduce the cross-section of a wire by pulling the wire through a single, or series of, drawing die(s). Lubrication in this drawing process is essential to obtain a good quality on the wire surface and also for maintaining good surface finish and long die life [1]. Various lubricants, such as oils [1-6], are employed, and its viscosity is the property closely related to its ability to reduce friction, in consequence viscosity affects the surface finish of wires, and the heat generated between wire and die(s). When the viscosity of a lubricant is too low, the lubricated wire will have inadequate protection and will therefore be subject to excessive wear [4-6]. When the viscosity of the lubricant is too high, the lubricated wire will expend additional energy to complete its task.

In cold-drawn process to obtain aluminum and copper wires commonly are used as lubricants mineral oils [4,5], however this conventional lubricants may have too high viscosity (internal friction that resists sliding and inhibits flow) to effectively slip between and lubricate the moving parts of these systems, particularly in newer truck components that are designed with close tolerances and tight fits. Conventional lubricants may also be heavy, making it harder for pumps, gears and shafts to move. These effects create energy losses and friction losses, and iste fuel.

On the other hand, low-viscosity lubricants are less resistant to flow than conventional lubricants, a property that helps reduce friction and energy losses. Depending upon the application, low-viscosity lubricants may also contain additives [3] designed to withstand the extreme pressure (EP) that could occur as the lubricant flows between the wire and die. Low-viscosity lubricants may be made from synthetic or mineral oil blends with low-viscosity and EP additives. Manufacturers generally offer low-viscosity blends as "fuel economy" lubricants, since the fuel-saving potential of these products is significant.

In this work, the rheology behavior of a mineral oil (conventional lubricant for cold-drawn of aluminum wire) and of an emulsion lubricant (conventional lubricant for cold-drawn process of copper wires) is compared with the rheology behavior of a new lubricant that can be used for both processes, cold-drawn process of aluminum and copper wires. To evaluate the efficiency of each lubricant studied, the roughness surface of each wire is measured by AFM. Results have indicated that, this new lubricant with a low viscosity that promotes a lower energy process also decreases the roughness of Al and Cu wires compared with conventional lubricants, i.e. it has a important influence in the quality of the surface wires. This means that this new lubricant could be used during the process of both metals without making important changes, which means low operations costs and flexibility for the manufacturing plant.

2. Experimental

2.1 Rheology behavior of lubricants

Three lubricants are analyzed by experimental measurements of viscosity vs shear rate. The first lubricant studied is a mineral oil; it is used as traditional lubricant for cold-drawn process to obtain Aluminum wire. The second one studied is an emulsion; it is used commonly as lubricant for the cold drawn-process to obtain copper wires. And finally, the third lubricant studied is a new emulsion that it is developed with the aim to be used for both cold-drawn process, Al and Cu.
The viscosity measurements are obtained by an Anton-Paar-Rheometer using geometry of concentric cylinders, see figures 1a and 1b.

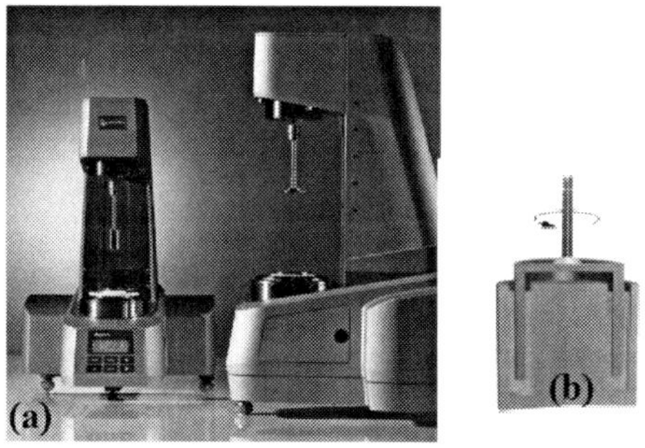

Figure 1, a) Anton Paar Rheometer Measuring Heat. b)The geometry of concentric cylinders

These lubricants are then used during the cold-drawn process to obtain aluminum and copper wires, obtaining four different samples: Al and Cu wires with their conventional lubricants and Al and Cu with this new lubricant common for both. Surfaces of these wires are analyzed by AFM in contact mode (see Figure 2) obtaining 3D-images. From these images the Root Mean Square (RMS) parameter is calculated for each AFM-image to evaluate the surface roughness.

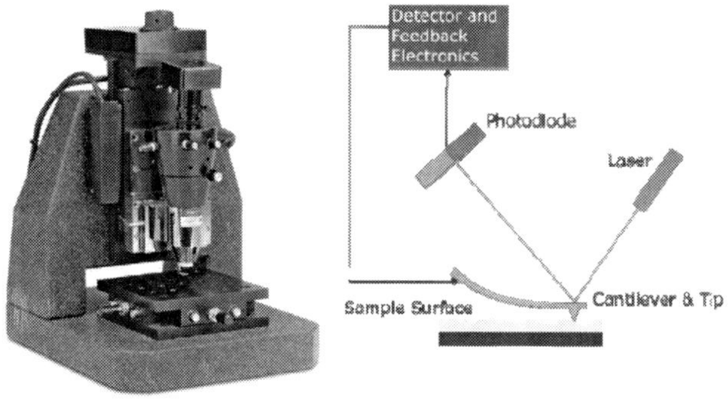

Figure 2. Atomic Force Microscopy

3. Results and discussion

First, all lubricants are characterized by FTIR technique, but like is a new product, these results will not be presented in this work, only a general discussion about this will be done. For mineral oil used as traditional lubricant in cold-drawn process to obtain aluminum wire, FTIR spectra display mainly vibrational modes for the following chemical groups: methyl and methylene groups. For the case of the emulsion used as traditional lubricant in the cold-drawn process to obtain copper wire, FTIR spectra display mainly vibrational modes for methyl and carbonyl groups. Finally for the new emulsion, FTIR spectra displays the same vibrational modes that are identified in the other samples, in addition a vibration mode that corresponds to hydroxyl group also is identified. These results suggest that the chemical structure for the new emulsion is more complex that chemical structure of traditional lubricants.

Experimental results obtained for the rheology behavior of lubricants studied are shown in Figure 3. All lubricants studied show constant viscosities for a shear rate variation, this phenomenon correspond to a Newtonian Fluid and it is a typical rheological behavior of several oils. In table 1 are reported the viscosity values for each lubricant.

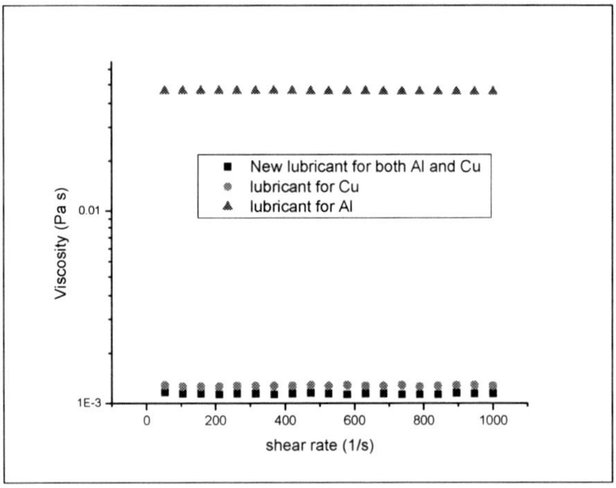

Figure 3. Rheology behavior of lubricants studied.

Table 1 resume of experimental results shown on figure 3.

Samples	**Viscosity (Pa s)**
Traditional lubricant for cold-drawn process to obtain copper wire	0.0021
New lubricant for both Al and Cu	0.0012
Mineral oil: Conventional lubricant for cold-drawn process to obtain aluminum wire	0.0419

The new emulsion has a lower viscosity (0.0012 Pa s) in consequence the cold-drawn process must expend lower energy to complete its task. However, the lubricated wire could have inadequate protection and will therefore be subject to excessive wear. To evaluate the protection of lubricant on the wires, roughness surfaces of Al and Cu wires are measured by AFM. In the next are presented results obtained for each wire studied.

3.1 Results of AFM analysis

Aluminum wire surfaces

Figures 4 show 3D-images of aluminum surfaces wires. Figure 4a corresponds to aluminum surface wire obtained from a cold drawn process using mineral oil (traditional lubricant for Al). Figure 4b corresponds to aluminum wire surface obtained using the new emulsion as lubricant in the cold drawn-process.

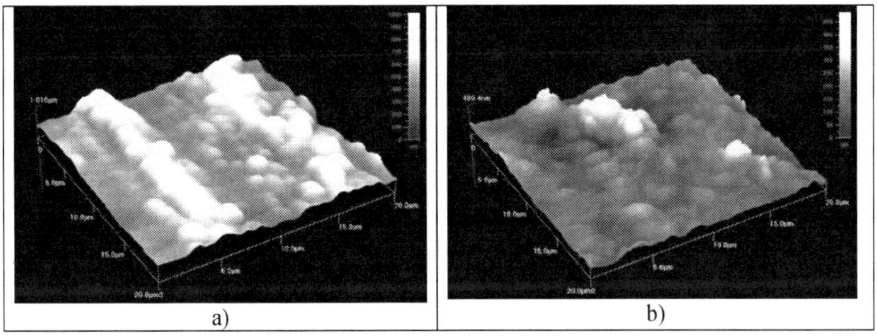

Figure 4. 3D-Images obtained by AFM for aluminum wire surfaces.

From these images is observed in a qualitative way that the roughness for figure 4b is lower than the roughness of figure 4a.

On the other hand, Figures 5 show 3D-images of copper surfaces wires. Figure 5a corresponds to copper surfaces wire obtained from a cold drawn process using the traditional lubricant (emulsion) for this process. Figure 5b corresponds to copper wire surface obtained using the new emulsion as lubricant in this cold drawn-process.

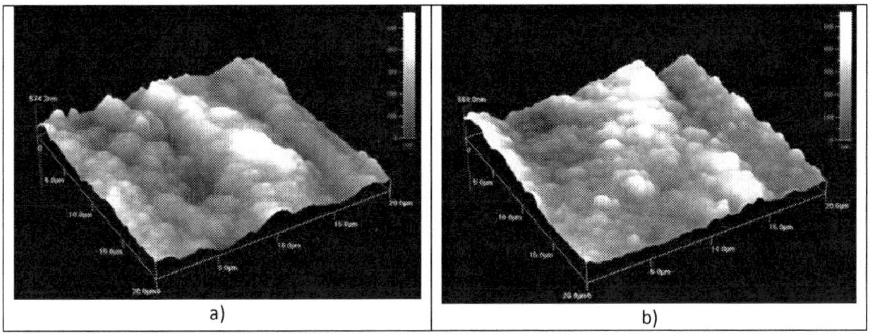

Figure 5. 3D-Images obtained by AFM for copper wire surfaces.

In this case, the difference of roughness between figure 5a and 5b is not as evident, as in the case of aluminum (figures 4a and 4b), for this reason it is necessary to calculate a roughness parameter (Root Mean Square or RMS) for each AFM-image. It must be noted that at least three different zones are characterized and this RMS value is the average of these three zones.

Table 2. RMS measurements obtained from AFM-images of 20x20 μm, for samples of aluminum and copper wires obtained by cold-drawn process, using in each case a different kind of lubricant.

Lubricants used	RMS measurements	
	Aluminum wires	Copper wires
Traditional lubricants	162.20 nm	101.06
New Emulsion	70.34 nm	72.87

From results shown in figures 4, figures 5 and Table 1, the new emulsion has a lower viscosity, and the roughness obtained for both cases aluminum and copper is lower than the roughness obtained in cold-drawn processes using the corresponding traditional lubricants. Moreover, other three different scanning size of 15x15, 10x10 and 5x5 μm are carried out also in three different zones, showing always lower RMS samples cold drawn with the new lubricant.

4. Conclusion

The new emulsion tested in this work is an important alternative to use in cold-drawn-processes for aluminum or copper wires. The roughness obtained in both cases is lower when the new emulsion is used as lubricant for each cold-drawn process improving the surface quality of the wires. And finally this new lubricant should improve the cost of the cold drawn process from the point of view of energy and time.

References

1. Ashbridge M.T.J, Gilmour K.R.; Leacock A.G. Proc. Instn. Mech. Engrs. 2001, Vol 215 Part J.
2. Lara Romero J., Chiñas Castillo F. , 2007, Vol. X, No. 34, pp. 16-23.
3. Miller D.J., Sun L., Walzak M.J., McIntyre N.S., Chvedov D., Rosenfeld A., Surf. Interface Anal. 2005, No.37, pp. 499–508.
4. Rameshkumar T., Rajendran I., Latha AD., Journal of Convergence in Engineering, Technology and Science, Vol.1, December 2009, pp. 9-16.
5. Imaizumi S., Igarashi M., Shibata J., Furukawa Review, 2005, No.28, pp. 39-45.
6. Vallés J.L., Vergara I.; "Tribología con nanomateriales", Investigación y Ciencia, 2008, pp. 48-57.

Mater. Res. Soc. Symp. Proc. Vol. 1373 © 2012 Materials Research Society
DOI: 10.1557/opl.2012.310

Steel-Magnesium Laminated Composites by Infiltration

Arda Çetin[1*], Jérôme Krebs[1], Alexandre Durussel[1, 2], Andreas Rossoll[1], Junya Inoue[2], Shoichi Nambu[2], Toshihiko Koseki[2], Andreas Mortensen[1]

[1] Ecole Polytechnique Fédérale de Lausanne (EPFL), Laboratory for Mechanical Metallurgy, CH-1015 Lausanne, Switzerland
[2] Department of Materials Engineering, The University of Tokyo
7-3-1 Hongo, Bunkyo-ku, Tokyo 113-8656, Japan

ABSTRACT

Steel–magnesium alloy laminated composites can be produced by gas-driven pressure infiltration of a molten magnesium alloy between layers of stacked steel sheets followed by directional solidification of the infiltrated magnesium alloy. A key step in the process is ensuring adequate separation and alignment of the steel sheets during the process; this is achieved by introducing small dimples in the steel sheets to hold them apart during infiltration. Advantages of the process are its speed, the defect-free composites it produces, and the fact that, unlike roll-bonded composites, the steel in the composite is in an annealed condition. The ultimate tensile strength of the as-cast laminates, of 260 MPa, obeys the rule of mixtures. The uniform tensile elongation, of around 20%, makes the infiltrated laminates nearly as ductile as the bulk steel it contains, implying that the magnesium alloy in the as-cast laminates has a substantially increased tensile ductility in comparison to the bulk state in a metallurgically equivalent condition.

INTRODUCTION

Lamination is a strategy, used well over a millennium, to drive certain metals and alloys to perform well beyond their performance range [1-5]. A brittle metal, for example, can experience large tensile strains when bonded to a ductile metal [6, 7] as long as the interfaces are strong enough to resist delamination. The reason is that, if the two metals are strongly bonded together, the more ductile metal will prevent necking in the other, less ductile, metal thus coaxing it to deform beyond the tensile strain that it would reach in its metallurgically equivalent bulk state [3-12].

Magnesium and its alloys tend to display poor ductility since they have only a few slip systems in their hexagonal crystal structure. Lamination could therefore be a promising strategy to improve their performance, notably where deformability is required. To this end, bonding magnesium to a more ductile metal, such as steel, can be attractive, also because magnesium is relatively inert in contact with iron. Chemical compatibility of various magnesium alloys with iron has already been investigated [5, 13, 14]; however, none of those studies concern laminated structures.

In what follows, we show how infiltration can be used to produce laminated metal composites (LMCs) of mild steel and a manganese and zinc containing magnesium alloy, and explore their structure and tensile properties. A more complete report of this work can be found in Ref. [1]

EXPERIMENT

The LMCs were made by infiltration of ZM21 magnesium alloy, a zinc and manganese containing magnesium alloy, into a stack of 0.1 and 0.2 mm thick ductile cold-rolled low-carbon steel (SPCC steel, Japanese standard JIS G3141). The elemental composition of the two constituent phases is given in Table 1.

Table 1. Elemental composition (wt %) of constituent materials.

	C	Si	Mn	P	S	Zn	Mg	Fe
SPCC Steel	0.04	0.01	0.017	0.013	0.010	-	-	Balance
ZM21 (nominal)	-	-	1	-	-	2	Balance	-

To prepare the steel stacks, the sheets were ground to clean and smoothen their surface and cut to desired dimensions of 28×100 mm^2 using automatic shears. The steel sheets were then punched to introduce small dimples on their surface, which keep them apart during infiltration. To this end, we use a small device including a punch, an upper and a lower aluminium plate with cylindrical holes 1.02 mm in diameter. The lower plate has matching conical indentations, 1 mm in diameter and 0.5 mm deep. The depth of the dimples is tailored to obtain equal layer thickness between alternating metal layers in resulting LMCs by altering the distance between the punch and the upper aluminum plate (0.1 mm and 0.2 mm, for LMCs with 0.1 and 0.2 mm layer thickness, respectively).

To test whether the dimples affect the mechanical response of the laminates, the LMCs were produced using two different dimple configurations. In the first configuration, identified as the *dimpled* configuration, some of the dimples were positioned in the gauge section of the laminates machined from the LMCs. In the other configuration, identified as the *nondimpled* configuration, the dimples were all positioned outside the gauge section of the LMCs.

Flat dog-bone tensile test samples (ASTM standard E8M) were machined from the infiltrated LMCs, with gauge dimensions of 25 mm × 6 mm × 2.2 mm. Tensile samples were also machined from a 1 mm thick SPCC sheet having undergone identical heat treatment with the infiltrated sheets and 2 mm thick cast ZM21 rectangular plates to measure the mechanical response of the constituent phases in their monolithic state. All tests were run on a Zwick (Ulm, Germany) screw-driven machine with a 10-kN load cell at a nominal strain rate of 6×10^{-4} s^{-1}.

Samples being pore free, the volume fraction of steel and magnesium alloy in each LMC were determined by densitometry, using a Sartorius MC 210P microbalance and a Sartorius Gravity Determination Kit YDK0.1 (IG Instrumenten-Gesellschaft, Zurich, Switzerland).

DISCUSSION

The microstructure of infiltrated LMCs are presented in Fig. 1 for LMCs with (a) 0.1 mm layer thickness and, (b) 0.2 mm layer thickness. The stacks are well infiltrated with reproducible layer thicknesses. The ZM21 alloy in the infiltrated LMCs display a very large grain size, on the order of millimeters. Statistically significant grain size and texture data were thus difficult to obtain, since each layer seems to be composed of only a few grains. The elongated grain structure of the SPCC steel, typical of cold-rolled steel recrystallizes into a fine-grained, equiaxed structure during infiltration (see Fig. 2).

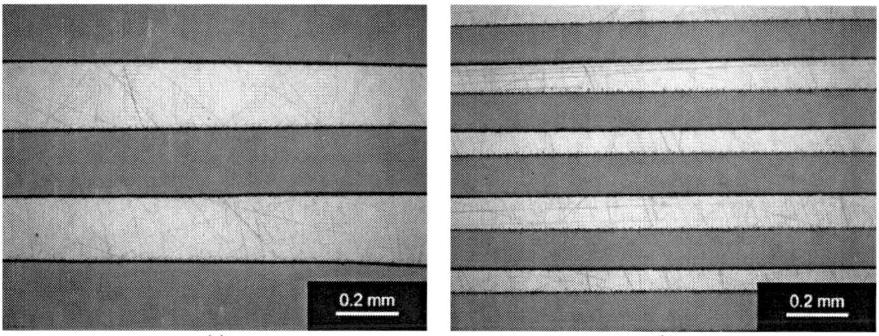

(a) (b)

Fig. 1 Optical micrographs of infiltrated LMCs with (a) 0.2 mm and (b) 0.1 mm layer thickness. Dark phase is steel and the light phase is ZM21.

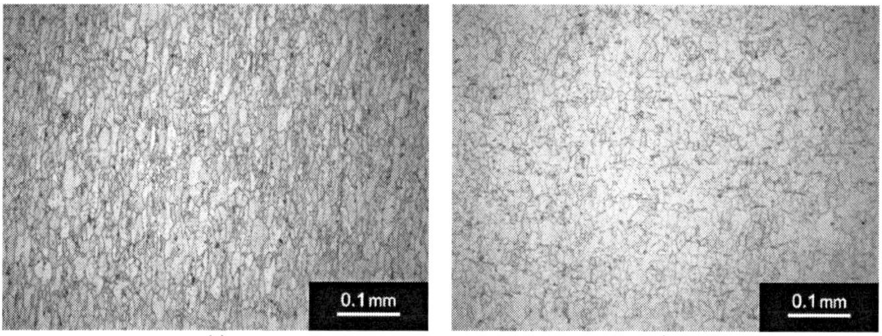

(a) (b)

Fig. 2 Grain structure of the SPCC steel (a) before and (b) after infiltration. The elongated grain structure of the steel recrystallizes into an equiaxed structure, with a mean size of 12 μm.

The interface is free of pores and/or any brittle intermetallic phases that would weaken it. EDX linescans (Fig. 3a) through the interface of the LMCs reveal that Mn, coming from the ZM21 alloy, segregates at the interface, with a penetration depth into the steel plates, on the order of a few μm. The remaining Mn in the Mg side precipitates as Mn-rich particles, roughly 1 μm size. Those particles are visible on the fracture surface of steel layers after tensile testing (Fig 3b).

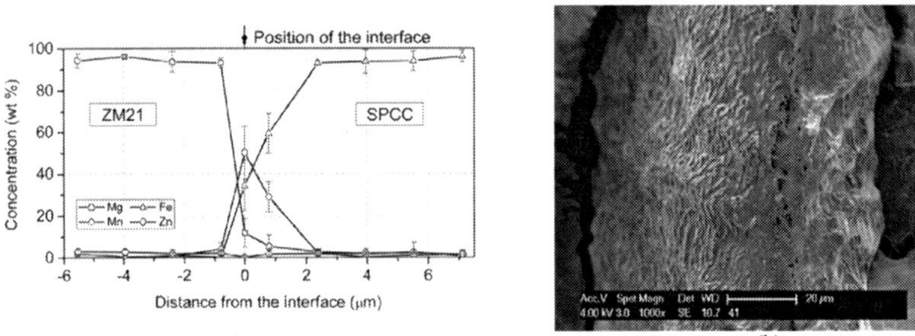

(a) (b)

Fig. 3 (a) Averaged line scans through the interface of a 0.2 mm LMC showing the Mn segregation at the interface, and (b) close-up of the fracture surface of a steel layer in a 0.2 mm LMC. The Mn particles, visible on both sides of the wedge, adhere to steel upon delamination.

The mechanical properties of the constituent phases are listed in Table 2, where E is the Young's modulus, $\sigma_{0.2}$ is the proof stress, UTS is the ultimate tensile strength and ε_u is the peak uniform strain. The mechanical properties of the LMCs are summarized in Table 3, where V_{SPCC} is the volume fraction of the steel in the LMCs, $\varepsilon_{u,th}$ is the theoretical uniform elongation of the composite, estimated as described below assuming composite deformation according to the equistrain rule of mixtures, and Z denotes the percent reduction in area at fracture in the neck. Two tensile tests were run for each condition.

Table 2. Mechanical properties of monolithic SPCC steel and ZM21 alloy.

	E [GPa]	$\sigma_{0.2}$ [MPa]	UTS [MPa]	ε_u [%]
SPCC steel	200	260	320	25.7
As-cast ZM21	45	131	225	13.7

Table 3. Tensile properties of infiltrated steel/magnesium alloy LMCs (by sample).

$t_{initial}$ [mm]	Dimpled	V_{SPCC} [%] ± 0.1%	$\sigma_{0.2}$ [MPa] ± 3 MPa	UTS [MPa] ± 3 MPa	ε_u [%] ± 0.7%	$\varepsilon_{u,th}$ [%]	Z [%] ± 1%
0.1	No	46.1	198	260	19.0	21.8	38
		46.6	200	259	17.9	21.9	38
	Yes	48.6	201	266	19.4	22.1	36
		48.1	183	265	24.1	22.0	40
0.2	No	52.5	209	263	22.4	22.5	42
		52.1	233	261	22.2	22.4	39
	Yes	50.4	198	264	17.2	22.3	28
		50.8	197	263	16.9	22.3	35

UTS values of all LMCs were found to be within 95 to 98% of rule of mixtures predictions (~260 MPa). A direct rule of mixtures estimation of the uniform strain of the laminate is, on the other hand, invalid [8-10]; one must therefore conduct a stability analysis on composite deformation, knowing that in the present instance the present composites deform according to the equistrain rule of mixtures. To this end, we use larger-strain power-law-fits [15] of the true-stress and true-strain curves of the two constituent phases to obtain for the SPCC steel sheet:

$$\sigma^{SPCC} = 595.3\ \varepsilon^{0.27} \qquad (1)$$

and for the ZM21 magnesium alloy:

$$\sigma^{ZM21} = 355.1\ \varepsilon^{0.16} \qquad (2)$$

Ignoring strain-rate effects, one can deduce the maximum uniform strain that can be attained by the composite by applying the Considere construction to the equistrain rule of mixtures LMC true-stress/true-strain curve computed on the basis of individual component flow laws, given in Eqs. (1) and (2). An example is illustrated in Fig. 4 for an LMC with 0.2 mm layer thickness. Uniform strain values of the LMC samples of this work, together with the equistrain rule of mixtures estimates for each particular case are reported in Table 3 as ε_u and $\varepsilon_{u,th}$, respectively. As can be seen, (*i*) in general the experimental values (ε_u) tend to be near, albeit on average somewhat lower than, predicted values ($\varepsilon_{u,th}$) and (*ii*) variations in experimental data exceed variations in $\varepsilon_{u,th}$ caused by the slight variations in V_{SPCC}.

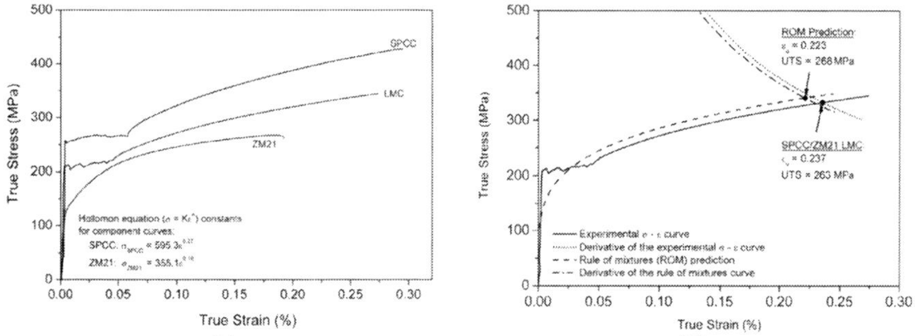

(a) (b)

Fig. 4 (a) Tensile curves of the component phases and the LMC, (b) illustration of the Considere constructions for the true stress/true strain behavior of the LMC, and the equistrain rule of mixtures prediction deduced by fitting constituent phase curves with Eqs. (1) and (2).

CONCLUSIONS

Steel-ZM21 alloy laminated composites with layer thickness of 0.2 and 0.1 mm can be produced by gas pressure infiltration. Dimpling the steel sheets to hold these apart during infiltration results in a neatly layered composite and not influence the mechanical response of the resulting LMCs up to peak tensile load, nor the ultimate tensile strength. The infiltrated LMCs are free of pores or intermetallic compounds and display uniform layer thickness. The ultimate tensile strength and the uniform elongation of the composites are only slightly lower than values given by the equistrain rule of mixtures.

ACKNOWLEDGMENTS

This work was sponsored by the Collaborative Research Program between the University of Tokyo and the Ecole Polytechnique Fédérale de Lausanne supported by the Toyota Motor Corporation.

REFERENCES

1. A. Çetin, J. Krebs, A. Durussel, A. Rossoll, J. Inoue, T. Koseki, S. Nambu, and A. Mortensen, *Metall. Mater. Trans. A*, DOI: 10.1007/s11661-011-0756-2.
2. C.S. Smith, *A History of Metallography*, 1988 paperback ed. (MIT Press, Chicago, IL, 1960), pp. 3-65.
3. J. Wadsworth and D.R. Lesuer, *Mater. Charact.* **45**, 289-313 (2000).
4. D.R. Lesuer, C.K. Syn, O.D. Sherby, J. Wadsworth, J.J. Lewandowski, and W.H. Hunt, Jr., *Int. Mater. Rev.* **41**, 169-67 (1996).
5. D. Embury and O. Bouaziz. *Annu. Rev. Mater. Res.* **40**, 213-41 (2010).
6. S. Nambu, M. Michiuchi, J. Inoue, and T. Koseki, *Compos. Sci. Technol.* **69**, 1936-41 (2009).
7. J. Inoue, S. Nambu, Y. Ishimoto, and T. Koseki, *Scripta Mater.* **59**, 1055-58 (2008).
8. S.T. Mileiko, *J. Mater. Sci.* **4**, 974-77 (1969).
9. S.L. Semiatin and H.R. Piehler, *Metall. Trans. A* **10A**, 85-96 (1979).
10. G. Garmong and R.B. Thompson, *Metall. Trans.* **4**, 863-73 (1973).
11. S. Nambu, M. Michiuchi, Y. Ishimoto, K. Asakura, J. Inoue, and T. Koseki, *Scripta Mater.* **60**, 221-24 (2009).
12. C.K. Syn, D.R. Lesuer, J. Wolfenstine, and O.D. Sherby, *Metall. Trans. A* **24A**, 1647-53 (1993).
13. J.C. Viala, D. Pierre, F. Bosselet, M. Peronnet, and J. Bouix, *Scripta Mater.* **40**, 1185-90 (1999).
14. M. Sacerdote-Peronnet, E. Guiot, F. Bosselet, O. Dezellus, D. Rouby, and J.C. Viala, *Mater. Sci. Eng. A* **445-446**, 296-301 (2007).
15. M. Selin, *Metall. Mater. Trans. A* **41A**, 2805-15 (2010).

Mater. Res. Soc. Symp. Proc. Vol. 1373 © 2012 Materials Research Society
DOI: 10.1557/opl.2012.311

Application of the intra-particle diffusion model for activated carbon fibers in an aqueous medium

Teresa Ramírez-Rodríguez[1] and Fray de Landa Castillo-Alvarado[1]
[1]Escuela Superior de Física y Matemáticas, Instituto Politécnico Nacional, 07738 Ave. IPN, D.F, México.

ABSTRACT

The intra-particle diffusion model (IPD), proposed by Weber and Morris has been applied to the analysis of the kinetics of adsorption on activated carbon fibers with phosphate groups in the removal of cadmium ions in aqueous media. It is evident that the removal of cadmium ion kinetic model of pseudo-second order provides a better fit than the model of pseudo-first order and the intra-particle diffusion model provides the best to the sample compared activating solution: grams fibers of 1:3.

INTRODUCTION

The removal of heavy pollutants is wastewater treatment has been studied by several researchers [1-3]. Some of the pollutants released in the wastewater are in relatively high concentrations, Ni, Zn, Cd, Pb and Cr. In this paper it has been consider the model intra-particle diffusion and adsorption kinetics to describe the uptake of cadmium ions in a porous adsorbent immersed in an aqueous solution. This process is considered to be controlled by the diffusion of soluble species through the pores in activated carbon fibers (ACF) and the simultaneous adsorption of the active sites within the walls of the pores.

THEORY

In the kinetic models have the hypothesis that the adsorption rate is controlled solely by the rate of the solute in the adsorbent surface and intra-particle diffusion and external mass transport cannot be considered. The adsorption kinetic is commonly modeled with the equations of pseudo-first and pseudo-second order.

Kinetic model of pseudo-first order

The adsorption rate equation of pseudo-first order was originally designed by Lagergren in 1898 for systems liquid-solid [4].

$$\frac{dq_t}{dt} = k\left(q_e - q_t\right) \qquad (1)$$

Where q_e y q_t (mg g^{-1}) are the adsorption capacities at equilibrium and at a time t, respectively (mg g^{-1}) and k, is the adsorption constant of pseudo-first order (min^{-1}). Integrating the equation (1) with boundary conditions: $t = 0$, $q_0 = 0$, $t = t_1$, $q_t = q_1$

It has obtained the integrated rate law for an adsorption reaction pseudo-first order, which becomes in its linear form:

$$\log(q_e - q_t) = \log(q_e) - \frac{k}{2.303}t \tag{2}$$

<u>Kinetic model of pseudo-second order</u>

In 1995 Ho presented the adsorption rate expression for pseudo-second order, which shows how the rate of adsorption depends on the adsorption capacity at equilibrium and the concentration of the adsorbate [5].

In an attempt to represent the equation of adsorption of divalent metals in a material during agitation was assumed that the process can be of second order and considering the chemical adsorption involving divalent metal covalent forces; second-order reaction may depend on the amount of divalent metal ions on the surface of the material and the amount of divalent metal ions adsorbed at equilibrium. The rate expression for adsorption is described:

$$\frac{dq_t}{dt} = k(q_e - q_t)^2 \tag{3}$$

Where k, is the rate constant of adsorption ($g\ mg^{-1}\ min^{-1}$), q_e is the amount of adsorbed divalent metal ions to balance ($mg\ g^{-1}$), and q_t is the amount of divalent metal ions on the surface of the adsorbent at any time t, ($mg\ g^{-1}$). Separating variables in equation (3):

$$\frac{dq_t}{(q_e - q_t)^2} = kdt \tag{4}$$

And integrating over the boundary conditions; $t = 0$, $q_0 = 0$ y $t = t_1$, $q_t = q_1$

$$q_t = \frac{q_e^2 kt}{1 + q_e kt} \tag{5}$$

Where equation (5) is the integrated rate law for a reaction of pseudo-adsorption of the second order, the equation in its linear form:

$$\frac{t}{q_t} = \frac{1}{h} + \frac{1}{q_e}t \tag{6}$$

$$h = kq_e^2 \tag{7}$$

Where h represent the initial adsorption rate (mg $g^{-1}min^{-1}$), when q_t/t approaches to 0. The model constants for the pseudo-second order can be determined experimentally by plotting t/q_t *vs* t.

Diffusion model intra-particle

The adsorption liquid/solid involves intra-particle diffusion and mass action. In physical adsorption, the mass action is a very fast and may not have greater significance for the kinetic study; therefore, the process of adsorption kinetics is controlled by diffusion in the film or by intra-particle diffusion, in a process should be rate-limiting step [6, 7].

The rate of adsorption is known is controlled by several factors including the following processes [8] (i) The diffusion of solute from the solution on the surface of the fiber, (ii) dissemination of the film to the surface of the particle (external diffusion), (iii) diffusion from the surface to internal sites (surface diffusion or pore diffusion) and (iv) uptake may involve several mechanisms, physical and chemical adsorption, ion exchange or precipitation. [9] Mackay and Poots noted that the fraction of solute adsorbed can be expressed in terms of the square root of time. A plot of the fraction of solute adsorbed on the square root of time can be used to estimate the rate of rate of intra-particle diffusion in the linear range. The mathematical dependence of the concentration in the solid $t^{1/2}$ has been deduced considering the adsorption mechanism to be controlled by diffusion in the adsorbent (as spherical particles) and the convective diffusion of the solution. The concentration varies with $t^{1/2}$ given by the equation.

$$q_t = k_p t^{1/2} + C \tag{8}$$

Where k_p is the slope of the graph (q_t *vs* $t^{1/2}$) which gives the initial rate of adsorption intra-particle diffusion controlled, $k_p(mg\ g^{-1}\ min^{1/2})$, the initial part of the curve is attributed to the limits of the effect of diffusion layer (which is the external film resistance).

The linear extrapolation of the straight lines with the time axis intercept gives C, which are proportional to the thickness of boundary layer [10].

The existence of diffusion process of the outer layer can be deduced from the fact that the graphs of the diffusion model intra-particle does not pass through the origin.

DISCUSSION

Six samples of ACF with phosphate groups in the removal of cadmium ions were subjected to a water bath at 30°C in a shaking 100rpm which maximum removal capacity was obtained at 240 minutes. In the activation of the solution in 2 ml of activating and grams of carbon fiber for 1:1, 1.5:1, 2:1, 3:1 and 1:2.

Kinetic model of pseudo-first order

The reaction rate values shown in figure1 that the concentration of cadmium ions in aqueous solution decreases in a time of 120 to 240min in all samples and a time interval of 240 to 360 minutes begins the desorption samples 1:1, 1.5:1, 2:1, 3:1 and 1:2, while 3:1 sample has a reduced concentration.

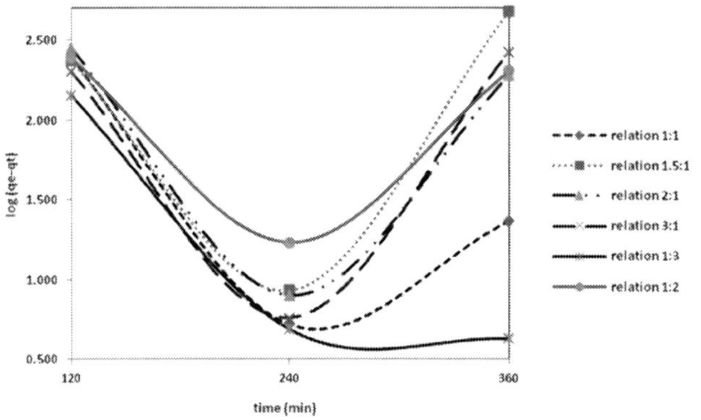

Figure 1. Kinetic model of pseudo-first order in the removal of Cd^{2+} with ACF

Kinetic model of pseudo-second order

Using the model of pseudo-second order has been obtained in figure 2 that the reaction rate in the samples 1:3, 3:1, 1:2 and 1:1 settings provide the best correlations, whereas samples 1.5:1 and 2:1 are those which have a less efficient performance.

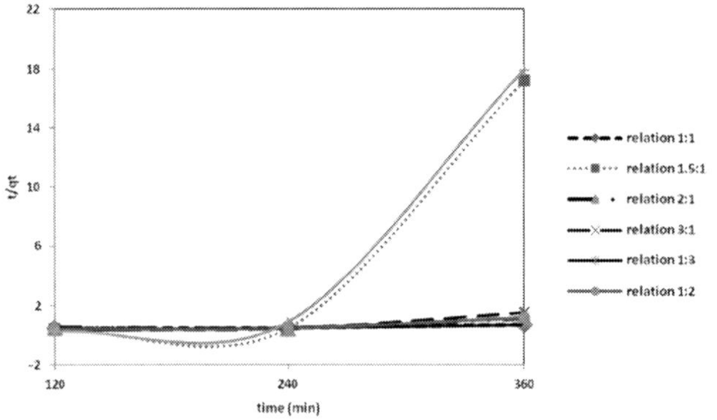

Figure 2. Kinetic model of pseudo-second order in the removal of Cd^{2+} with ACF

Diffusion model intra-particle

The figure 3 by applying the diffusion model indicates that intra-particle samples with ratio of 1:1, 1.5:1, 2:1, 3:1 and 1:2, optimal diffusion occurs at a $t^{1/2}$ averaged 15.49$min^{1/2}$, while diffusion occurs in the ratio 1:3 in a $t^{1/2}$, 10.9544$min^{1/2}$.

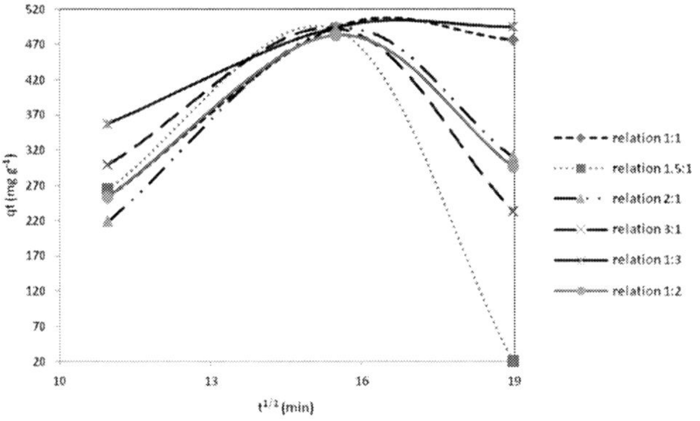

Figure 3. Diffusion model intra-particle in the removal of Cd^{2+} with ACF

Table I contain the results of the analysis of the kinetics of adsorption of Cd^{2+} model using pseudo-first order, pseudo-second order and intra-particle diffusion.

Table I. Adsorption kinetic models $Cd^{1/2}$

Relations	Model of pseudo-first order			Model of pseudo-second order				Diffusion model intra-particle	
	K_{ad} [min^{-1}]	q_e [mg g^{-1}]	R^2	q_e [mg g^{-1}]	h [mg g^{-1} min^{-1}]	k [g $mg^{-1}$$min^{-1}$]	R^2	k_p [mg g^{-1}]	R^2
1:1	-0.01083	330.44560	0.3696	833.33300	3.436	4.94E-06	0.7800	29.168	0.7600
1.5:1	0.00299	48.43950	0.0275	14.34700	-9.40E-02	-0.00045	0.7500	-26.446	0.2000
2:1	-0.0016121	112.0211	0.0103	400	8.1699	5.10E-05	0.6800	13.903	0.1600
3:1	0.00161	50.76910	0.0044	212.76600	-3.021	-0.00007	0.8100	5.6874	0.0285
1:3	-0.01474	480.39660	0.7790	625.00000	8.006	-6.67E+00	0.9800	17.873	0.8156
1:2	-0.00069	114.57760	0.0041	322.581	76.336	0.00073	0.7800	7.557	0.0624

It has been obtained that the adsorption kinetics model of pseudo-second order provides the best fir correlation model compared with pseudo-first order, whereas in the diffusion model intra-particle samples 1:3 and 1:1 have the best correlation.

CONCLUSIONS

It has been analyzed with kinetic models of activated carbon fibers with phosphate groups, including the kinetic model of pseudo-second order correlation shows better results than 0.75 this being mostly about the kinetic model of pseudo-first order. On the other hand, model intra-particle diffusion is concluded that relations activating solution: fiber, 1:1, 1.5:1, 2:1, 3:1, 1:2 in the optimum occurs at a $t^{1/2}$ average of 15.49min$^{1/2}$, while diffusion occurs in the ratio 1:3 in a $t^{1/2}$ 10.9544min$^{1/2}$, which provides the best performance in the diffusion on the adsorbent.

REFERENCES

1. S. Gupta, D. Kumar, J. P. Gaur, "Kinetic and isotherm modeling of lead (II) sorption onto some waste plant materials" *Chem Engineering J.* **148**, 226 (2009).
2. O. P. Raul, L. R. Roberto, M. B. Jovita, G. C. Rosa M., "Adsorption rate of phenol from aqueous solution onto organobentonite: Surface diffusion and kinetic models" *J. of Colloid and Interface Sci.* **364**, 195 (2011).
3. R. Ocampo-Perez, R. Leyva-Ramos, P. Alonso-Davila, J. Rivera-Utrilla, M. Sanchez-Polo, "Modeling adsorption rate of pyridine onto granular activated carbon" *Chem, Eng. J.* **165**, 133 (2010).
4. Y. S. Ho, G. McKay, "Kinetic models for the sorption of dye from aqueous solution by wood" *Institution of Chemical Engineers*, **76B,** 183 (1998).
5. Y.S, Ho, C. C. Wang. "Pseudo isotherms for the sorption of cadmium ion onto tree fern" *Process Biotec,* **39,** 759(2004**).**
6. K. S. Hui, C. Y. H. Chao, S. C. Kot, "Removal of mixed heavy metal ions in wastewater by zeolite 4ª and residual products from recycled coal fly ash" *J. of Hazard. Mater.* **B127**, 89 (2005).
7. M. D. Pritzker, "Model for parallel surface and pore diffusion of an adsorbate in a spherical adsorbent particle", *Chem. Engineering Sci.* **58**, 473 (2003).
8. A. Findon, G. McKay, H. S. Blair. "Transport studies for the sorption of cooper ions by chitosan". *J Environ Sci Health.* **A28,** 173 (1993).
9. G. McKay, V. J. Poots. "Kinetics and diffusion processes in colour removal from effluent using wood as an adsorbent" *J Chem Technol Biotech.* **30,** 279 (1980).
10. J.C. Igwe, C.O. Onyegbado, A. A. Abia. "Studies on the kinetics and intraparticle diffusivities of BOD, colour and TSS reduction from palm oil mill effluent (POME) using boiler fly ash". *African J. of Environ. Sc. and Techn.* **4(6),** 392 (2010).

Mater. Res. Soc. Symp. Proc. Vol. 1373 © 2012 Materials Research Society
DOI: 10.1557/opl.2012.312

Experimental Study on Mixing in Gas-Stirred Ladles with and without the Slag Phase through a Water Physical Model

Adrián M. Amaro-Villeda, Jorge A. González Bello, Marco A. Ramírez-Argáez.
[1]Facultad de Química, Universidad Nacional Autónoma de México, Mexico City, Mexico
adrianvilleda@yahoo.com.mx, marco.ramirez@.unam.mx

ABSTRACT

A 1/6th gas–stirred water physical model of a 140 ton steel ladle is used to evaluate mixing in air–water and air–water–oil systems to model argon–steel and argon–steel–slag systems respectively. Thickness of the slag layer is kept constant at 0.004 m. The effect of the gas flow rate (7, 17, and 37 l/min), plug position (0, 1/3, ½, and 2/3 of the ladle radius, R), and number of plugs (1, 2, and 3) on mixing time is also analyzed in this work. Gas is injected at the bottom of the ladle under several plug configurations varying both position and number of plugs. Chemical uniformity of 95% is selected as mixing criterion. Mixing times are experimentally determined when a tracer is suddenly injected into the ladle and the model is instrumented with a pH meter to track the time evolution of the tracer concentration (NaOH 1 M solution) in a given location inside the ladle. Process conditions for best mixing in both water–gas and water-gas–slag systems are: a single plug located at 2/3 of the ladle radius with a gas flow rate of 17 l/min.

INTRODUCTION

There is growing interest to improve steel properties through optimization of steelmaking processes. Ladle furnaces play a key role in promoting steel quality where refining operations such as deoxidation, desulphurization, inclusion removal and/or modification, adjustment of the chemical composition, thermal and chemical uniformity are performed. These refining operations are assisted and accelerated by stirring the liquid steel by argon gas injected through porous plugs located at the ladle bottom and in the presence of a slag layer on the top of the steel. Although this process has been widely used in steelmaking for several decades, there are several physical phenomena not fully understood in secondary refining, which drives numerous research efforts currently being developed [1-3]. Research tools include physical and mathematical modeling.

Measurements in physical models have characterized the gas-liquid plume structure, gas holdup, bubble frequency, bubble velocity, mixing times, heat and mass transfer.

Mathematical models, developed by stating the set of conservation equations involved in the gas-liquid fluid flow problem, have been used to represent hydrodynamics of the ladle, heat and mass transfer as well as mixing and turbulent structure in the ladle.

There are many studies in gas-stirred ladles by physical modeling on mixing phenomena which do not present a general agreement about the process conditions required to optimize mixing and these works do not take into account the slag layer [4-9]. Therefore, mixing studies in ladles are still necessary in ladle systems with the presence of the slag.

In this work a study on mixing in a gas-stirred ladle assisted by an acrylic physical model is performed by injecting a tracer into the physical model filled with water, a layer of oil on top of the bath and air injected at the bottom, simulating steel, slag and argon respectively. The objective is to determine the effect of gas flow rate, location and number of plugs on mixing time in a three phase (steel-slag-gas) physical model of a gas-stirred ladle.

EXPERIMENTAL PROCEDURE

An acrylic physical model is filled with distilled water and a layer of red engine oil on its top to simulate steel and slag, respectively. Air is injected from the bottom of the ladle through nozzles and the flow rate coming from a gas compressor is controlled by flowmeters (see Figure 1). Mixing time is monitored by measuring the bath pH with a probe located at a fixed position inside the ladle. Mixing time is determined as the time taken to reach 95% chemical uniformity starting from the time at which a NaOH 1M solution is injected at the top of the ladle. Experiments are recorded with a video camera and data of NaOH concentration and time are casted in a Microsoft Excel™ worksheet. Table I presents dimensions of the physical model which is scaled to 1/6th of an industrial ladle.

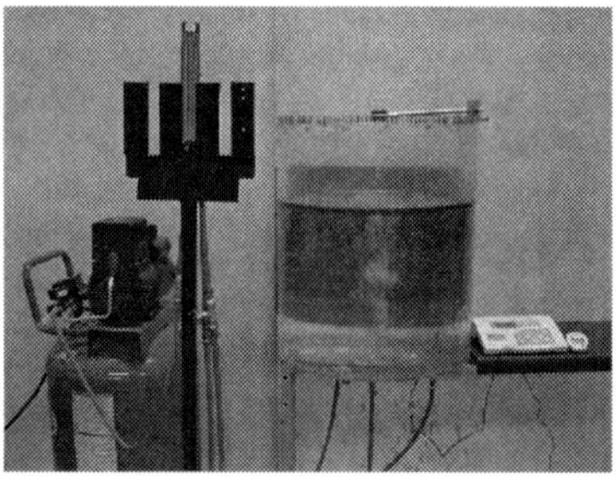

Figure 1. Photograph of the experimental setup.

Table I: Geometric similarity between our physical model and the prototype

	Height of furnace ladle (m)	Bath height (m)	Furnace diameter (m)
Industrial ladle (140 Tons)	3.736	2.460	3.226
Physical Model	0.622	0.410	0.537

In addition to geometric similarity with a scale factor λ=1/6, dynamic similarity is accomplished by scaling the forces driving the flow by equating the modified Froude numbers in both the model and the industrial ladle. This dynamic scaling determines gas flow rates to be injected into the model (Q_{mod}) as a function of gas flow rates used in industrial ladles (Q_{fs}).

Equation 1 presents the relationship between Q_{mod} and Q_{fs} while Table II shows flows computed from Equation 1 and used in the physical model.

$$Q_{\text{mod}} = \lambda^{2.5} Q_{fs} \tag{1}$$

Table II: Data for scaling gas flow rate in the physical model

	Gas flow rates (m^3/min)			Gas flow rates (l/min)		
Industrial ladle	0.882	1.764	3.527	882	1764	3527
Physical model	0.007	0.017	0.037	7	17	37

Table III shows steel and water physical properties. As can be seen, steel and water have practically the same kinematic viscosity, which results in kinematic similarity and warrants similar motions for both the water and the steel. This similarity explains why water has been widely used to simulate liquid steel in physical modeling.

Table III: Physical properties of water and steel

	Density (ρ, Kg/m^3)	Molecular viscosity (η, N s^2/m^2)	Kinematic Viscosity (ν, m^2/s)
Water at 20 °C	998.2[10]	0.0010[10]	$1X10^{-6}$
Steel at 1600 °C	7000[11]	0.0068[12]	$0.97X10^{-6}$

An experimental design accounted for the number of plugs (1, 2 and 3 plugs), plug position (r/R= 0, 1/3, 1/2 y 2/3) and air flow rate (7, 17, 37 l/min) as the main process variables. Figure 2 shows several gas injection configurations formed by changing the number of plugs and their position at the bottom of the ladle. Internal nozzle diameter is 0.005m. All experiments in the experimental design are run with and without slag layer (slag layer of 0.004m).

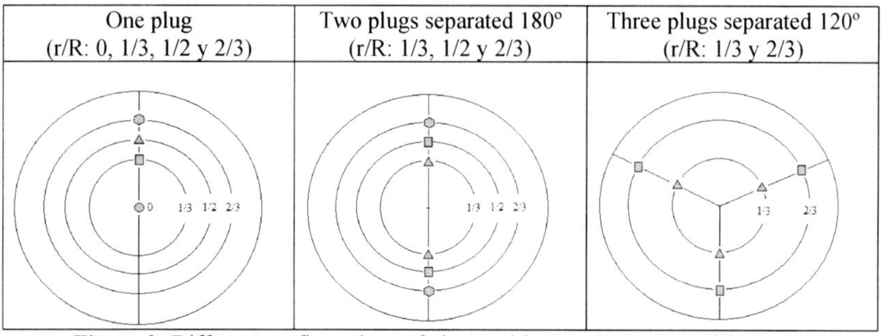

Figure 2: Different configurations of plug positions at the bottom of the ladle.

RESULTS AND DISCUSSION

Figure 3(a) shows the tracer evolution with time for a gas flow rate of 17 l/min injected with one plug located at 2/3R. Every experiment is run 3 times to determine the experimental error and therefore three lines are included in this figure. Additionally, three horizontal lines can be seen which represent the equilibrium concentration and concentrations 5% below and above the equilibrium, i. e. the mixing criterion of 95% of uniformity. From this type of plots, mixing times are determined when the tracer concentration meets the mixing criteria. Figure 3(b) shows mixing times as a function of plug position, with one plug without oil and a gas flow rate of 7 l/min. As the plug is moved away from the center towards the wall, mixing time decreases (this result is also found for air flow rates of 17 and 37 l/min). As the plug moves from the center to the wall, better stirred baths are created by a single circulation flow pattern of liquid, covering larger ladle volumes, while center gas injection creates two upper circulations but the bottom of the ladle presents dead zones. Presence of slag (oil) increases mixing times since water losses momentum by friction with the oil phase.

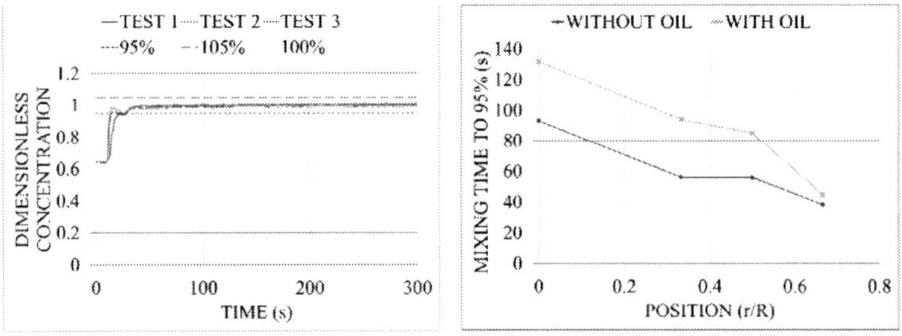

(a) (b)

Figure 3: (a) Dimensionless concentration vs. time to determine mixing time with one plug located at r/R=2/3 and a gas flow rate of 17l/min with oil. (b) Mixing time as a function of plug position, with and without a layer of oil, with one plug and a gas flow rate of 7 l/min.

To evaluate the effects of gas flow rate and number of plugs, plug location of 2/3 of radius is fixed (best mixing plug position). Figure 4(a) shows mixing times versus gas flow rate for a single plug at r/R=2/3, with and without oil. Mixing time decreases with an increment in gas flow rate since more bubbles drag more fluid and create more turbulence. Additionally, poorer mixing is achieved with oil than without oil due to the friction at the water oil interface.

Figure 4(b) shows the mixing time as a function of the gas flow rate with three plugs at r/R=2/3, with and without oil. It can be seen that mixing is worst with more plugs at a constant gas flow rate. In this case, every plug creates an independent circuit of fluid which does not allow the tracer injected inside a liquid loop to reach the other loops by convection and makes turbulent transfer the only mechanism available to reach the entire ladle. Again, mixing is improved without the oil layer and with an increment in the gas flow rate.

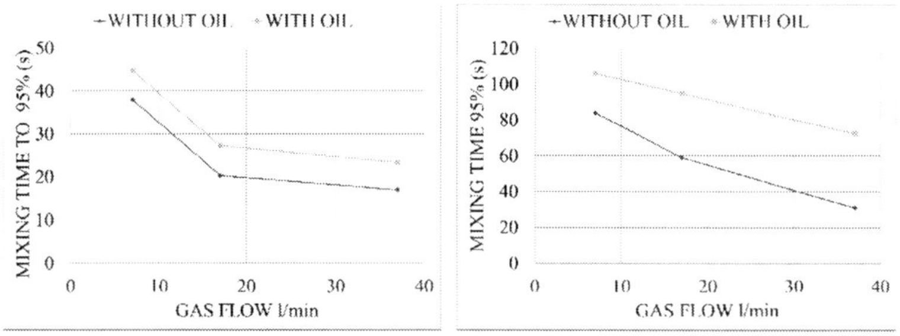

(a) (b)

Figure 4: (a) Mixing time as a function of gas flow rate with and without oil, with one plug at r/R=2/3. (b) Mixing time as a function of gas flow rate with and without oil, with three plugs separated 120° and located at r/R=2/3.

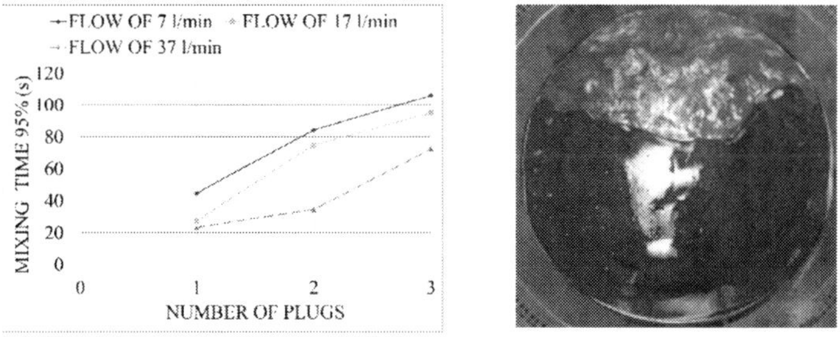

(a) (b)

Figure 5: (a) Mixing time with oil varying the number of plugs and air flow rate. (b) Top view of the physical model with an oil layer of 0.004 m, with one plug at r/R=2/3, and the air flow rate of 7 l/min.

Table IV: Estimated averaged area of water exposed with one, two and three plugs at r/R=2/3, varying the air flow rate

	Number of plugs	Air flow rate (l/min)		
		7	17	37
Percentage of exposed bath area	1	30.26	46.97	67.44
	2	35.46	51.62	59.86
	3	33.15	46.15	63.04

Figure 5(a) shows mixing times as a function of the number of plugs located at R= 2/3 with oil. As the number of plugs increase mixing times increase. Each plug creates an independent recirculation flow pattern and the tracer injected above one plug is entrapped in one loop and it cannot reach the other loops by convection making turbulent transfer the only mechanism to reach mixing.

Figure 5(b) shows a photograph depicting the exposed surface area ("open eyes") created by the gas plumes. The case presented showed the lowest percentage of exposed area (two plugs located at 2/3R and a gas flow rate of 7 l/min). Table IV shows percentage of area exposed for the different plug configurations explored in this work. Less area exposed is better since steel is not over exposed to air preventing oxidation.

CONCLUSIONS

Blowing gas away from ladle centre, with and without oil, decreases the mixing time due to convective motion promoted in most of liquid volume.

Mixing time decreases with increasing the gas flow rate, with and without oil, since as air flow increases more drag is exerted by bubbles to liquid, enhancing convection and turbulence.

Mixing times are increased in the presence of a layer of oil on the bath due to the momentum transfer or friction between water and oil.

Mixing time increases in the oil-water system by increasing the number of plugs, due to development of independent recirculation circuits, preventing convective mass transport. Under these conditions, turbulent mass transport becomes the only way to reach chemical equilibrium.

Lower mixing times with oil are found for a single plug at 2/3R and gas flows of 17 and 37 l/min (27.33 and 23.33 seconds respectively).

REFERENCES

1. D. Mazumdar and R. Guthrie: The physical and mathematical modelling of gas stirred ladle Systems, ISIJ int. **35** No.1, 1-20 (1995).
2. D. Mazumdar and J. W. Evans: Macroscopic models for gas stirred ladles, ISIJ Int. **44** No.3, 447 (2004).
3. P. G. Jönsson, and L. T. I. Jonsson: The use of fundamental process models in studying ladle refining operations, ISIJ Int. **41**, 1289 (2001).
4. M.-Y. Zhu, T. Inomoto, I. Sawada: Fluid flow mixing phenomena in the ladle stirred by argon through multi-tuyere, ISIJ int. **35** No.5, 472-479 (1995).
5. Y. Takatsuka y M. Iguchi: Effect of swirl motion on mixing time in water bath agitated by upward gas injection, ISIJ int., **41**, No.2, 124-127 (2001).
6. D. Mazumdar, H. B. Kim and R. I. L. Guthrie: Modelling criteria for flow simulation in gas stirred ladles: experymental study. Ironmaking & Steelmaking, **27**, No.4, 302 (2000).
7. M. Madan, D. Satish, and D. Mazumdar. Modeling of mixing in ladles fitted with dual plugs, ISIJ Int., **45**, 677. (2005).
8. J. Mandal, S. Patil, M. Madan and D. Mazumdar: Mixing time and correlation for ladles stirred with dual porous plugs, Metall. Trans., **36B**, 479 (2005).
9. J. W. Han, S. H. Heo, D. H. Kam, B. D. You J. J. Pak and H. S. Song: Transient fluid flow phenomena in a gas stirred liquid bath with top oil layer-approach by numerical simulation and water model experiments, ISIJ Int., **41** No. 10, 1165-1173 (2001).

Mater. Res. Soc. Symp. Proc. Vol. 1373 © 2012 Materials Research Society
DOI: 10.1557/opl.2012.313

Particle Swarm Optimization for Milling Titanium Alloy

I. Escamilla[1], L. Torres[1], B. Gonzalez[1], P. Zambrano[1]
[1] Facultad de Ingeniería Mecánica y Eléctrica. Ave. Universidad s/n. San Nicolás de los Garza, N. L. indira.escamillasl@uanl.edu.mx

ABSTRACT

Optimum machining parameters are of great concern in manufacturing environments, where economy of machining operation plays a key role in competitiveness in the market. Many researchers have dealt with the optimization of machining parameters for milling operations. In this paper, optimization procedures based on particle swarm optimization algorithm are developed for find machining parameters in milling operation. It describes development and utilization of the methodology that determines optimum Pareto's front analyzing feed, speed and depth for milling operation. The relationships between machining parameters and the performance measures of interest are obtained by using experimental data and a swarm intelligent neural network system. Results show that particle swarm optimization is an effective method for solving multi-objective optimization problems, and also, that an integrated system of neural networks and swarm intelligence can be used to solve complex machining optimization problems.

Keywords: Multi-objective optimization, particle swarm optimization, machining parameters, Titanium Alloy

INTRODUCTION

Determining and optimizing the parameters involved in a machining process is a critical and very important task. Machining of Titanium into finished components is receiving increasing attention in especially aerospace and automotive industries. Titanium is also one of the most expensive metals to produce, which is another drawback to manufacturing parts using this metal. Surface finish is an important parameter in the machining process. Surface quality is one of the most specified customer requirements. A major indication of surface quality on machined parts is surface roughness. Thus machinists and companies specializing in the machining of titanium materials generally develop techniques to maximize surface integrity of titanium alloys. Thus optimum properties usually are achieved during the production machining of titanium. [1, 2].

Neural networks (NNs) are computational intelligent methods, which can establish a mapping through the numerical inputs and outputs. The NNs extract a sensitive and realistic relation from some experimental input–output data, called training set. In this way, they can interpolate synthetic data that estimate the results of the experiments that have not been established, and predict optimum reasonable processing conditions [3].

Particle Swam Optimization (PSO) is an evolutionary computation (EC) method inspired by flocking birds. This population based stochastic optimization technique has been developed by Eberhart and Kennedy [4] and applied to many different systems including machining. Coello and Lechuga [5] find that PSO is an efficient alternative over other stochastic and population-based search algorithms, especially when dealing with multi-objective optimization problems after testing with several benchmark problems. The

PSO is an efficient method to locate the Pareto front of the multi-objective optimization problems; it is relatively easy to implement and has fewer parameters to adjust compared to generic algorithms. PSO is initialized with a population of random solutions and this initial population evolves over generations to find optima. However, in PSO, each particle in population has a velocity, the velocity vector reflects the socially exchanged information, which enables them to fly through the problem space instead of dying and mutating. Therefore, each particle is represented by a position and a velocity. Modification of the position of a particle is performed by using its previous position information and its current velocity. Each particle knows its best position (personal best) so far and the best position achieved in the group (group best) among all personal bests , in general, is defined in the following equations (1) and (2). Where are random values. Particles tend to be influenced by the success of anyone they are connected to.

$$\bar{x}_i = \bar{x}_i(t-1) + \bar{v}_i(t) \tag{1}$$

$$\bar{v}_i(t) = W\bar{v}_i(t-1) + C_1 r_1(\bar{x}pbest_i - \bar{x}_i(t)) + C_2 r_2(\bar{x}pbest_i - \bar{x}_i(t)) \tag{2}$$

Surface roughness depends on different machining parameters and its prediction and control is a challenge to the research. Owing to advances in computing power, there is an increase in the demand for the use of intelligent techniques. Recently, researches are directed toward hybridization of intelligent techniques. For example Jesuthanam et al. [6] propose a hybrid NNs trained with Genetic Algorithm (GA). Ozel et al. [7] propose a comparison between NNs and Regression.

EXPERIMENTAL STUDY

First we determine the parameters to be taken into account for the cut i.e., the ranges for the depth of machining, the feed rate and speed, which are the input of the neural network. The machining is done in a Vertical Machining Center Bridgeport VMC 760 (Figure 1a), using rectangular pieces of titanium (Ti 6Al-4V) of size 125 x 47 x 22 mm, the tool is an endmill coated with Aluminum Titanium Nitride (AlTiN) with 4 cutting edges and 3/8" diameter. The temperature is measured with Fluke 574 infrared thermometers shown in Figure 1b. The milling is carried out over a length of 47 mm, using a matrix with 3 factors and 3 levels (see Table 1), giving a total of 27 experiments with 3 runs.

Table 1: Machining parameters for the test

Condition	Units	Level 1	Level 2	Level 3
Speed	**m/min**	70	80	90
Feed	**mm/rev**	0.11	0.13	0.15
Depth	**mm**	0.50	1.00	1.50

(a) (b)

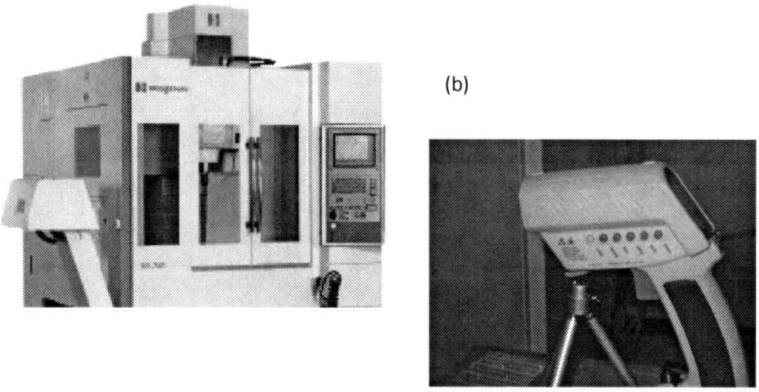

Figure 1. (a) Vertical Machining Center VMC 760. (b) Fluke 574 infrared thermometers

In this investigation, we use a three-factor full factorial design to determine the effects of speed, feed and depth of cut on temperature and roughness surface in milling. We performed three replicates of each factor level combination in order to account for variability in the process. Table 2 shows some temperature values out of 27 parameter combinations obtained in the tests. They are used to make a prediction model that represents the machine under study. The main purpose of this study is to determine the best parameters for the machining of the titanium alloy 6AL-4V and also optimizing the temperature at the same time.

Table 2. Selection of obtained results.

	Design of Exp			Roughness Replicates			Mean Temperature Replicates		
test	**Speed**	**Feed**	**Depth**	**1 Ra**	**2 Ra**	**3 Ra**	**1 Temp.**	**2 Temp.**	**3 Temp.**
1	1	1	1	0.8669	0.8472	0.8794	40.6	38.2	43.0
2	1	1	2	0.8442	0.8586	0.8644	45.5	57.7	44.2
3	1	1	3	0.9383	0.852	0.9043	61.9	63.2	54.8
4	1	2	1	0.8188	0.8796	0.8265	38.3	39.1	39.3
5	1	2	2	0.9369	0.8541	0.8512	48.7	50.3	46.2

RESULTS AND DISCUSSION

<u>Neural Network Back propagation.</u>

It is important to identify the following parameters in order to build the network. (1) The set of training patterns, input and target, (2) a value for learning rate, (3) a criterion that terminates the algorithm, (4) a methodology to update weights (5) the nonlinearity function,

(6) the initial weight values and (7) the learning moments. To develop this research a multilayer perceptron with a back propagation learning rule is used. Some of the variables in use are: Tinp = Neurons of the input layer and bias, Tmid = Neurons of the hidden layer, Tout = Neurons of the output layer, eta = Learning Constant, alpha = Moment. The parameter values are obtained by designing the experiment. Tables 3 and 4 show the best results obtained in the design of experiment, they will be used to train the network.

Table 3. Results with the best values for the variables used in the Neural Network for the roughness

Tinp	Tmid	Tout	Eta	Alpha	Netpochs	Error4
3 + 1	34	1	0.95	0.52	15000	3.6708e-4

Table 4. Results with the best values for the variables used in the Neural Network for the temperature

Tinp	Tmid	Tout	Eta	Alpha	Netpochs	Error4
3 + 1	36	1	0.8	0.65	11000	3.7492e-005

The input parameters considered in the neural network are speed, feed and depth of cut, and the output value is the surface roughness and the temperature. After performing different tests to train the network, results can be obtained by using the best neural network parameters. They show that the network output data are reliable and close to the real data. The neural network is trained with 80% of the values, the results are shown in Fig 2a to the Roughness and Fig 2b to the Temperature. A validation test is shown in the figure 3a for the Roughness. Figure 3b shows the validation test as a function of temperature

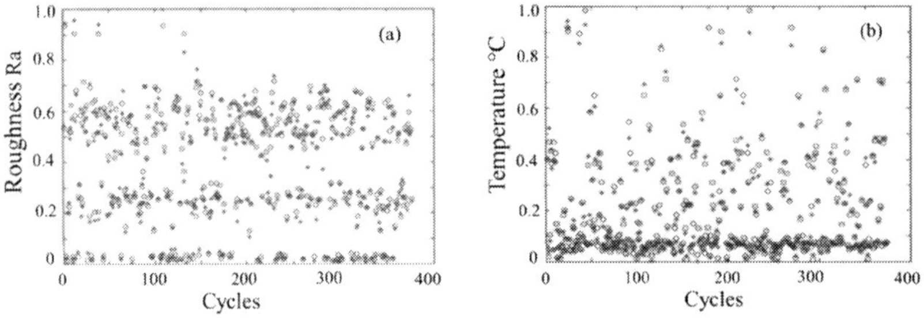

Figure 2. Experimental values as a function of trained data obtained from Neural Network Back Propagation. (a) Roughness, (b) Temperature. ◆ Real O Prediction.

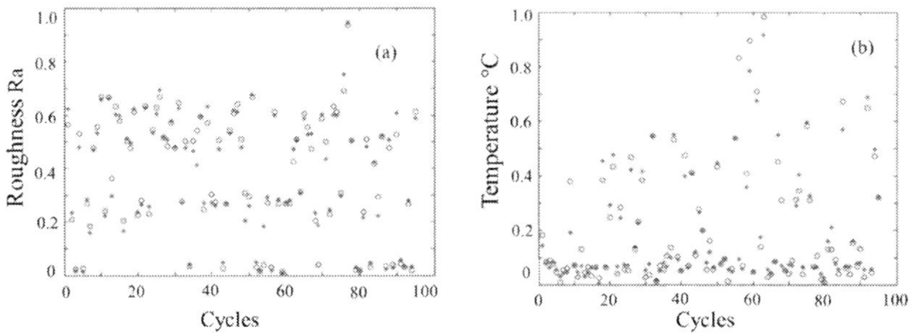

Figure 3. Validation test (a) For Roughness, (b) For Temperature. ◈ Real ○ Prediction

SIMULATION RESULTS

For each data set, experiments are implemented to minimize the influence of random effects. Each experiment uses a different randomly generated initial population. In addition, there are some parameters in the multi objective PSO which need to be specified by the user as shown in table 4.

Table 4.- Variable values used in the PSO

Alpha	beta1	beta2
0.25	0.25	0.25

The outcome of this algorithm is a Pareto set of optimal solutions to improve the machining parameters. Figure 4 shows the initial possible values (machining parameters) as dispersed dots as well as the predicted optimized results of the multiobjective PSO algorithm i.e., diamonds in a quiasi linear tendency at the bottom of Fig 4. This demonstrates that PSO has the capability to optimize temperature and roughness at the same time for the milling process under analysis. Figure 5 is a magnified view of the Pareto solution.

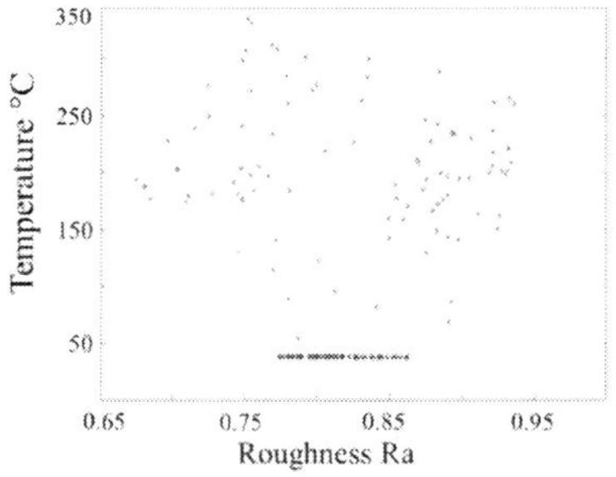

Figure 4. Initial data and Pareto front for for temperature and roughness.
◈ Real ◆ Prediction.

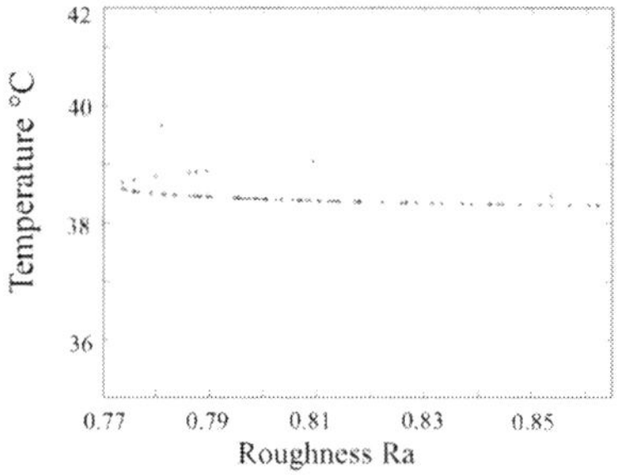

Figure 5. Magnified view of pareto front for temperature and roughness.
◈ Real ◆ Prediction.

CONCLUSION AND FUTURE WORK

In this paper, we introduce a procedure to formulate and solve optimization problems for multiple objectives that can exist in milling processes. We use neural network modeling together with particle swarm optimization technique. The representative multiple objectives for milling are defined to obtain a group of optimal process parameters, which minimizes temperature and surface roughness values at the same time. The neural network models integrated with the particle swarm optimizer called swarm-intelligent neural networks system can be determined in order to obtain a family of solutions that provides useful information to the user during the selection of machining parameters. Therefore, Pareto optimal fronts are computed and presented with stem plots for representing machining parameters yielding to a certain merit of interest such as surface roughness, temperature that can be selected by the user according to production requirements. The results indicate that the proposed swarm intelligent approach for solving the multiobjective optimization problem with different objectives is both effective and efficient, and can provide information in production planning for multi-parameter milling processes. The present work has considered only a two dimensional problem, temperature and surface roughness.

REFERENCES

1. Brunette J, Jeu-Herault M, Songmene V, Masounave J (2004). Mach Sci Technol 8(1):141–170
2. Davim JP, Silva LR, Festas A, Abrao AM (2009). Mater Des 30:228–234
3. Caydas U, Hascalik A (2008) J Mater Process Technology 202(1–3):574–582
4. R.C. Eberhart, J. Kennedy, Piscataway, NJ, 1995, pp. 1942-1948.
5. Coello C.A. and M.S.Lechuga, (2002) Proceedings of IEEE, World Congress on Computational Intelligence, 2002, pp.1051-1056
6. Jesuthanam CP, Kumanan S, Asokan P (2007). Mach Sci Technology 11(2):271–286
7. Ozel T, Karpat Y (2005). Int J Mach Tools Manuf 45(4–5):467–479

Mater. Res. Soc. Symp. Proc. Vol. 1373 © 2012 Materials Research Society
DOI: 10.1557/opl.2012.314

Multi-Objective Optimization of Corrosion Rate Parameters in Refining Process Using Particle Swarm Optimization.

B. González[1,2], L. Torres[2], F. A. Reyes[1], I. Escamilla[2], C. Vera[1], R. Colas[2]
[1]Corporación Mexicana de Investigación en Materiales, S.A. de C.V., Ciencia y Tecnología # 790, Fracc. Saltillo 400, Saltillo, Coahuila, México, C.P. 25290.
[2]Facultad de Ingeniería Mecánica y Eléctrica., Universidad Autónoma de Nuevo León., Ave. Universidad s/n., San Nicolás de los Garza, N. L., México, C.P. 66450

ABSTRACT.

In this paper, a Particle Swarm Optimization (PSO) algorithm is presented to find the optimal combination of corrosion rate parameters for a refining process in the oil industry. The experimental data in this paper are constituted by results obtained from field tests. Maintenance control is a very important aspect in order to prevent substantial damage to facilities, equipment and people. Other important factor to consider is the cost of maintenance which tends to reduce the required actions. The main parameters in corrosion control are flow, concentration of sulfur species, total acid number (TAN), temperature, and chromium content. However it is not easy to know the combined effect of different variables due to synergistic effects. Particle swarm optimization (PSO) is a population based stochastic optimization technique, inspired by social behavior of bird flocking or fish schooling. The system is initialized with a population of random solutions and searches for optima by updating generations. In PSO, the potential solutions, called particles, fly through the problem space by following the current optimum particles.

INTRODUCTION.

The corrosion of refinery equipment during oil distillation has been first noticed as early as 1920 [1]. As the oil producing and refining industries have further developed, this problem broadens and has become one of the most crucial points. Experience has shown that the main sites of corrosion attack are the components of oil pumping and refinery transfer lines, such as pipelines, valves and gates, heat exchangers, pipe stills, bubble sections, hydrocarbon stock feeders, fractionation-tower reflux units, etc. [2]. It causes severe damage to industrial equipment and materials. The adoption of preventive measures that can reduce or eliminate corrosion is financially costly and time consuming. In fact, corrosion prevention and treatment consume more than 20% of a typical industrial budget. Accidents related to corrosion processes resulted in equipment outage and economic losses, thus making it necessary to study the causes and the nature of this type of corrosion. Petroleum naphthenic acids (PNA) present in crude oils produced in many world regions are considered now to be the main factor responsible for the corrosion problem [1–8]. At oil processing temperatures, PNA show corrosion activity, inducing the specific corrosion type termed "naphthenic acid corrosion" (NAC) in the literature. In addition, sulfur dioxide, which is a component of both crude oil and natural gas, with its corrosive characteristics, undoubtedly enhances the corrosion rate of the equipment used in the production and extraction of oil. Ever since, people have established a series of models on corrosion, such as empirical formula, numerical simulation, general linear model, neural network model, etc. This study is an attempt to investigate corrosive behavior by PSO algorithm in improvement of corrosion rate in refining process. The variables of potential importance in this problem are material alloy, temperature and mixture of crude oil [9]. The cost of crude oil represents approximately 80% of the costs incurred by refineries. Naphthenic acid corrosion and sulphidation however can present serious threats to plant components with conventional metallurgy. Crude oils with high contents of organic acids, mainly naphthenic acids, are corrosive towards carbon steels at temperatures >200°C (392°F), which generate a demand for stainless steels containing molybdenum. Corrosion resistance is a function of unit design, operating temperature and flow velocity, as well as of the presence of compounds with a passivating action in crude oil. Difficulties in performing experiments and the low reproducibility of results, even in the laboratory, have led to the need to deal with naphthenic acid corrosion through the application of thermodynamic considerations and modeling [10].

Particle Swarm Optimization.

In order to establish a common terminology, in the following we provide some definitions of several technical terms commonly used: *Swarm*: Population of the algorithm. *Particle*: Member (individual) of the swarm. Each particle represents a potential solution to the problem being solved. The position of a particle is determined by the solution it currently represents. *pbest* (personal best): Personal best position of a given particle, so far. That is, the position of the particle that has provided the greatest success (measured in terms of a scalar value analogous to the fitness adopted in evolutionary algorithms). *lbest* (local best): Position of the best particle member of the neighborhood of a given particle. *gbest* (global best): Position of the best particle of the entire swarm. *Leader*: Particle that is used to guide another particle towards better regions of the search space. *Velocity* (vector): This vector drives the optimization process, that is, it determines the direction in which a particle needs to "fly" (move), in order to improve its current position. Inertia weight: Denoted by W, the inertia weight is employed to control the impact of the previous history of velocities on the current velocity of a given particle. Learning factor: Represents the attraction that a particle has toward either its own success or that of its neighbors. Two are the learning factors used: C1 and C2. C1 is the cognitive learning factor and represents the attraction that a particle has toward its own success. C2 is the social learning factor and represents the attraction that a particle has toward the success of its neighbors. Both, C1 and C2, are usually defined as constants. Neighborhood topology: Determines the set of particles that contribute to the calculation of the lbest value of a given particle [11].

PSO is an evolutionary computation (EC) method inspired by flocking birds. This population based stochastic optimization technique developed by Eberhart and Kennedy [12] and applied to many different systems. Coello and Lechuga [13] find the PSO is an efficient alternative over other stochastic and population-based search algorithms, especially when dealing with multi-objective optimization problems after testing with several benchmark problems. The PSO is an efficient method to locate the Pareto front of the multi-objective optimization problems; it is relatively easy to implement and has fewer parameters to adjust compared to genetic algorithms. PSO is initialized with a population of random solutions and this initial population evolves over generations to find optima. However, in PSO, each particle in population has a velocity; the velocity vector reflects the socially exchanged information, which enables them to fly through the problem space instead of dying and mutations. Therefore, each particle is represented by a position and a velocity. Modification of the position of a particle is performed by using its previous position information and its current velocity. Each particle knows its best position (personal best) so far and the best position achieved in the group (group best) among all personal bests. , in general, is defined in the following equations (1) and (2).

Where r_1, $r_2 \in [0,1]$ are random values. Particles tend to be influenced by the success of anyone they are connected to.

$$\vec{x}_1(t)=\vec{x}_1(t-1)+\vec{v}_1(t) \quad (1)$$

$$\vec{v}_1(t)=W\,\vec{v}_1(t-1)+C_1r_1(\vec{x}pbest_i-\vec{x}_1(t))+C_2r_2(\vec{x}pbest_i-\vec{x}_1(t)) \quad (2)$$

EXPERIMENT.

In the past, various methods have been used to quantify the impact of corrosion parameters on refining heavy crude oils. Though the processes that previous researchers have utilized are similar in nature, they all vary slightly in their execution. All of the relevant literature includes some kind of design of experiments that allows for a systematic approach to quantifying the effects of a finite number of parameters.

The aim of this study is to predict, as exactly as possible, the corrosion of stainless steel material when crude oil is processed in atmospheric distillation or vacuum towers and in this way obtain the longest lifetime of components in these processes. For this purpose we use field data from refinery plant (see Table I).

Table I. Parameters used for the test.

Parameter	Units	Values
oil mixture	(Heavy/Light)	1,2,3,4,5
temperature	(°C)	250,300,350,400
materials	(% Cr)	5,9,12,18

RESULTS AND DISCUSSION.

An analysis of variance (ANOVA) to determine the differences in corrosion rate using Minitab®. For the corrosion rate generation, all the factors are apparently significant in Table II. The regression analysis technique using least squares estimation is applied to get the coefficients of the model with the experimental data. figure 1 shows the regression model considering all variables including the interactions and the model generated parameters, analyzing them the inflation values are bad that means a poor performance, already that the coefficient of regression of the variance were estimated by the least squares method and are very inflated, resulting in having the largest VIF, this gives us very unstable estimates indicating that magnitudes and signs that can change greatly from one sample to another, but the R Square are fine.

Table II. ANOVA.

SOURCE	DF	MS	F	P-value
MIXTURE	4	2.4554	228.38	0.000
TEMP	3	62.0512	5771.53	0.000
MATERIAL	3	2.2485	209.14	0.000
MIXTURE*TEMP	12	1.4068	130.85	0.000
MIXTURE*MATERIAL	12	0.4294	39.94	0.000
TEMP*MATERIAL	9	0.2422	22.53	0.000
MIXTURE*TEMP*MATERIAL	36	0.3287	30.57	0.000
Error	80	0.0108		
Total	159			

Regression Analysis : corr rate versus Mixture, Temp., ...

The regression equation is

corr rate = 4.22 + 0.633 Mixture - 0.0451 Temp + 0.105 Matls - 0.0127 Mix Sqr+ 0.000112 Temp Sqr - 0.00424 Matls Sqr - 0.00230 Mixture-Temp+ 0.0100 Mixture-Matls- 0.000155 Temp-Matls - 0.000025 Mix-Temp-Matls

Predictor	Coef	SE Coef	T	P	VIF
Constant	4.223	2.851	1.48	0.143	
Mixture	0.6334	0.5839	1.08	0.282	247.7
Temp	-0.04509	0.01473	-3.06	0.003	246.3
Matls	0.1052	0.1654	0.64	0.527	223.4
Mix Sqr	-0.01272	0.03136	-0.41	0.686	26.7
Temp Sqr	0.0001125	0.00002099	5.36	0.000	212.3
Matls Sqr	-0.00424	0.002684	-1.58	0.119	33.0
Mixture-Temp	-0.002303	0.001676	-1.37	0.174	250.6
Mixture-Matls	0.00996	0.04615	0.22	0.830	378.5
Temp-Matls	-0.000155	0.0004641	-0.33	0.739	221.0
Mix-Temp-Matls	-0.0000248	0.0001399	-0.18	0.860	402.7

S= 0.469345 R-Sq =87.3% R -Sq (adj) = 85.4%

PRESS = 20.6656 R - Sq(pred) = 82.72%

Figure 1. Full square polynomial

A comparison of prediction results is shown in figure 2 between the full square polynomial regression model and the actual data. These prediction results are obtained from the full quadratic regression model by introducing each of the main parameters in Table 1.

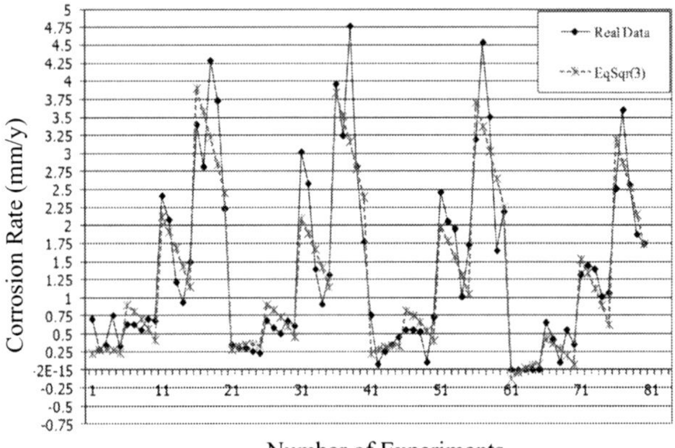

Figure 2. Data comparison between full square polynomial regression model and real data.

In the PSO algorithm the objective is obtain the best set of parameters by using two or more functions, in this case we take into account two objectives. In the first case, equation (3) has been considered to predict the corrosion rate from the full square polynomial regression model. The results are then compared by analyzing the corrosion rate as a function of chromium content and changing only the value according to eq. (4). As a result, plots are built by using both parameters as shown in figure 3. In this figure, the dots show the first population of optimal solutions to form the pareto front and to minimize the corrosion rate.

$$y1=4.22+(0.633*x1)-(0.0451*x2)+(0.105*x3)-(0.0127*x1^2)+(0.000112*x2^2)-(0.00424*x3^2)-(0.00230*x1*x2)+(0.01*x1*x3)-(0.000155*x2*x3)-(0.000025*x1*x2*x3) \quad (3)$$

Where: $x1$= oil mixture (Heavy/Light): 1,2,3,4,5
$x2$= temperature (°C): 250,300,350,400
$x3$= materials (%Cr): 5,9,12,18

$$y2 = 5, 9, 12, 18 \text{ (\% chrome)} \quad (4)$$

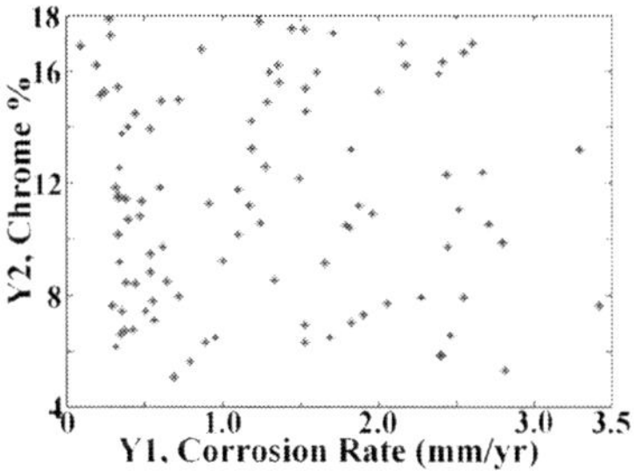

Figure 3. First particles movement in PSO.

As seen in the figures 4 and 5, these are the second and third populations of optimal solutions to form the pareto front and the particle movement to minimize the corrosion rate i.e., to obtain the mentioned objectives by evaluating the results and identifying the dominating values that give rise to the pareto front.

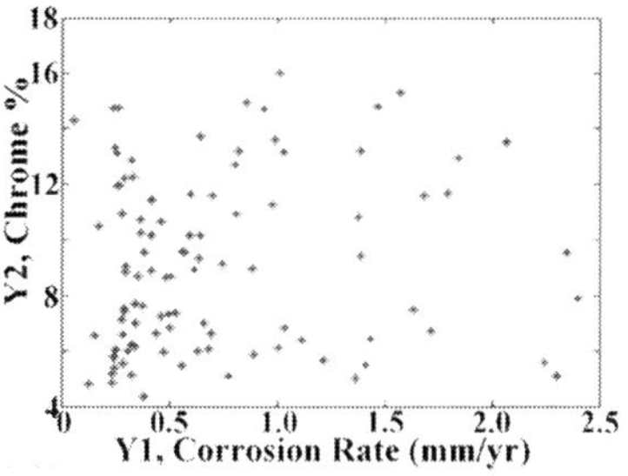

Figure 4. Particles movement.

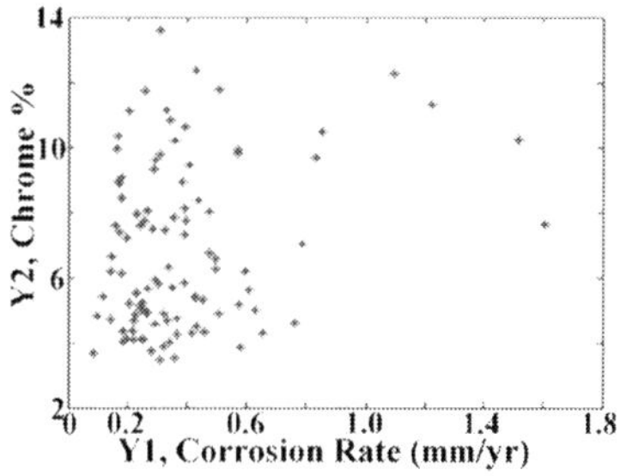

Figure 5. Particle movement in next position.

The outcome of this algorithm is a pareto set of optimal solutions to improve the corrosion rate (*), like those shown in figure 6 (see asterisks). This shows that the PSO algorithm has the capability to optimize the chromium content and the corrosion rate at the same time for this refining process. The points with circles are the first solutions found with the algorithm (○).

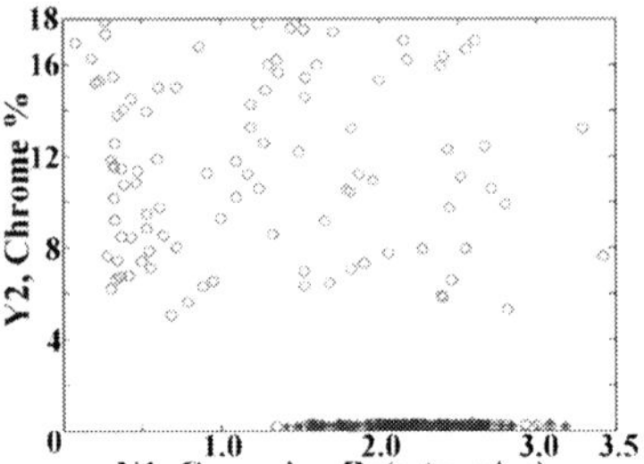

Figure 6. Initial data(○) and pareto front(*).

CONCLUSIONS.

This analysis can also be performed on other oil processes. It can consider other materials, vary the temperature, and analyze flow rates simply by using a neural network with the characteristics of the same historical data. The neural network models integrated with the particle swarm optimizer called swarm-intelligent neural networks system is then proposed in order to obtain a family of solutions that provides useful information to the user during the optimization of corrosion rate. The actual model proposed is thus useful for the interpretation of the process; however, the prediction is risky to be considered because the high VIF values. There are other alternatives to improve the confidence level, like the use of ridge regression or robust regression, but this is considered as a future work in this investigation. The present work has considered only a two dimensional problem by using corrosion rate and chrome percentage. In the future, a three dimensional parameter set will be considered.

REFERENCES.

1. A. Jayaraman, H. Singh, and Y. Lefebvre, Rev. Inst. Fr. Pet.**41**, (1986) 265.
2. E. Slavcheva, B. Shone, and A. Turnbull, Br. Corros. J. **34**, (1999) 125.
3. C. M. Cooper, Hydrocarbon Process. **51**, (1972) 75.
4. J. Gutzeit, Oil Gas J. **74**, (1976)156.
5. J. Gutzeit, Mater. Perform. **16**, (1977) 24.
6. E. Babaian-Kibala, H. L. Craig, G. L. Rusk, et al., Mater. Perform. **32**, (1993) 50.
7. E. Babaian-Kibala, H. L. Craig, G. L. Rusk, et al., in *Proceedings of the Conference "Corrosion'93", New Orleans, 1993*, p. 631.
8. E. Slavcheva, B. Shone, and A. Turnbull, Corrosion **54**, (1994) 922.
9. Hajeeh, M.. Estimating corrosion: a statistical approach. *Materials and Design, 24:* (2003)*509–518.*
10. S.P. Trasatti, G. Gabetta,. Study of naphtenic acid corrosion by neural network.
11. Margarita Reyes-Sierra and Carlos A. Coello Coello. (2006) Int. J. Computational Intelligence Research. Vol.2, (2006) pp. 287–308.
12. R.C. Eberhart, J. Kennedy, , Proceedings of IEEE International Conference on Neural Networks, Piscataway, NJ, 1995 pp. 1942-1948.
13. Coello C.A. and M.S.Lechuga, Proceedings of IEEE, World Congress on Computational Intelligence (2002) pp.1051-1056.

Mater. Res. Soc. Symp. Proc. Vol. 1373 © 2012 Materials Research Society
DOI: 10.1557/opl.2012.315

Molecular perspective of MoS_2/Co unsupported catalyst using computer assisted TEM simulations

Manuel Ramos[1,2], Joseph Serrano-Matos[5], Eduan Martinez-Soto[5], Brenda Torres[2], Gilles Berhault[4], Domingo Ferrer [3], José Rurik Farias-Mancilla[1], Russell R. Chianelli[2]
[1]Departamento de Física y Matemáticas, IIT –UACJ Cuidad Juárez, México, 32310.
[2]Materials Research and Technology Institute, UT-El Paso, El Paso TX.
[3]Microelectronics Research Laboratory, UT-Austin, Austin TX.
[4]IRCELYON, CNRS – University of Lyon, Villeurbanne, 69100, France.
[5]Departamento de Química, Universidad Metropolitana, San Juan, PR.

*Corresponding author: 450 Ave. Del Charro, Cd. Juárez, Chihuahua. IIT-UACJ, Edificio G Cubículo-A301. E-mail address: manuel.ramos@uacj.mx, maramos1@utep.edu

ABSTRACT

Understanding the morphology of catalytically active materials has been approached successfully in past decades using field electron microscopy in scanning and transmission modes. In this respect, some simulated TEM measurements for unsupported promoted molybdenum di-sulfide (MoS_2/Co) provided some insights about molecular structure in those catalytic layered transition metal sulfides (LTMS). However, due to poor resolution, lack of color enhancement, and other factors, sections of those materials observed under TEM do not resolve the structure by itself; in particular about the localization of cobalt atoms for promoted MoS_2 unsupported catalyst. This work concludes an epitaxial growth of MoS_2 slabs over (111)-Co_9S_8 crystallographic plane, with a stacking degree size of 6 slabs. Results presented in here are obtained using experimental HRTEM and TEM simulations using the multi-slice method with a slice thickness of 25 Å and projected potential $f(U) = \sum_{i=1}^{n} a_i e^{(-b_i U^2)}$, where a_i and b_i are coefficients to be determined.

Key words: HRTEM, MoS_2, Catalytic, Projected Potential

INTRODUCTION

Transition Metal Sulfides like MoS_2 have been used in the past twenty years as catalytic materials for HDS reactions in crude oil [1], the catalytic activity on layered MoS_2 can be increased when it is promoted with nickel or cobalt [2] due to the Sabatier principle [3]. In the past decade extensive morphology and chemical activity studies has been conducted with much success using transmission and scanning electron microscopy, powder X-ray diffraction and Mössbauer spectroscopy [4-8]. However, important information has not been brought to literature. It has been applied as a scientific tool to describe two main factors (1) the role of promoters (i.e. Co, Ni) and (2) Structure/Function of physical-chemical properties. Commercial software and open access computational quantum code has made possible to model some of this layered materials, with good success [9, 10]. Here we present a series of images produced by a combination of experimental HRTEM and TEM simulations [11, 12]; this particular study provides particular insights of the layered MoS_2/Co nanostructure.

EXPERIMENT

Samples are prepared by mixing ammonium tetrathiomolybdate, $(NH_4)_2MoS_4$ (ATM) with $CoCl_2$ in aqueous solution. Typically, 1.6 g of $(NH_4)_2MoS_4$ in 20 mL of water is added under stirring at room temperature to a second solution of 0.47 g of $CoCl_2.1/2\ H_2O$ dissolved into 10 mL of water leading to the formation of a precipitate, the obtained Co/Mo molar ratio is 0.3 [13]. Experimental images are obtained using a Schottky field emission electron microscope FEI Tecnai TF20 (200 kV), under Scherzer defocus conditions, that occurs at $\Delta f_{Sch} = -1.2\ (C_s\lambda)^{1/2}$, equipped with an STEM high-angle annular dark-field (HAADF) detector, and X-Twin lenses with a resolution of ~ 2 nm. Samples for HRTEM observations are prepared by adding 0.05 g of catalytic powder in ethanol in a glass vial for ultrasonic particle dispersion and placed in a 200 mesh C/Cu grid. The series of microstructure reconstructions are produced with the aid of Accerlys Cerius[2] package. The superlattice of MoS_2 slabs are manipulated manually to produce bending effects and interaction with 111-Co_9S_8 surface, everything is placed in a super-crystal (P1 non-symmetry) with a = 102 Å, b = 67Å and c = 20Å , $\alpha = \beta = \gamma = 90^o$. In order to produce experimental HRTEM image reconstruction, a multislice TEM simulator is implemented with a and b coefficients ranging from 84 to 512 and slice thickness of 25 Å. Fitting of images is done electronically as described in chart flow presented in figure 1.

RESULTS AND DISCUSSION

An experimental HRTEM image is presented in figure 2a, follow by its molecular model as constructed using DFT package 2b. Then in order to confirm experimental observation a simulation of HRTEM is done using multislice method with a & b values of 256, fig.2c. Then finally a digital matching between experimental HRTEM and molecular model is produced in order to see fitting parameters in an aim of resolved the structure, fig. 2d. While doing the image interpretation we are able to locate six 2H-MoS_2 slabs interacting with (111)-Co_9S_8 structure in a (101)-(111) planar match. High surface tension is generated during nucleation/growth since slabs slightly bend as observed in experimental and simulated HRTEM images.

This nucleation seems to be occurring because of the formation of the Co_9S_8 phase first during hydrothermal reaction. Further examination in other sections of this catalytic structure, we are able to observe a dislocation of MoS_2 slabs as presented in figure 3; the dislocation is occurred because slabs are chemically attach by weak van der Waals, allowing to move easy right before have any contact with (111)-Co_9S_8 plane [16]. The fundamental understanding of interaction, activation and location of catalytic activity at molecular-crystal level, is first proposed by Delmon and co-workers in their synergic contact [14, 15], which states absorption of DBT molecules are generated in the interface between this two crystals. Figure 3 presents a schematic of Delmon synergic contact (taken from ref. 16).

Using the latest advances in transmission electron microscopy instrumentation, we are presenting here a new way of understanding this synergic contact model, with a description of the observed nanostructures between MoS_2 and Co_9S_8, that are responsible for hydrogenation of DBT molecules during HDS reaction of crude oil.

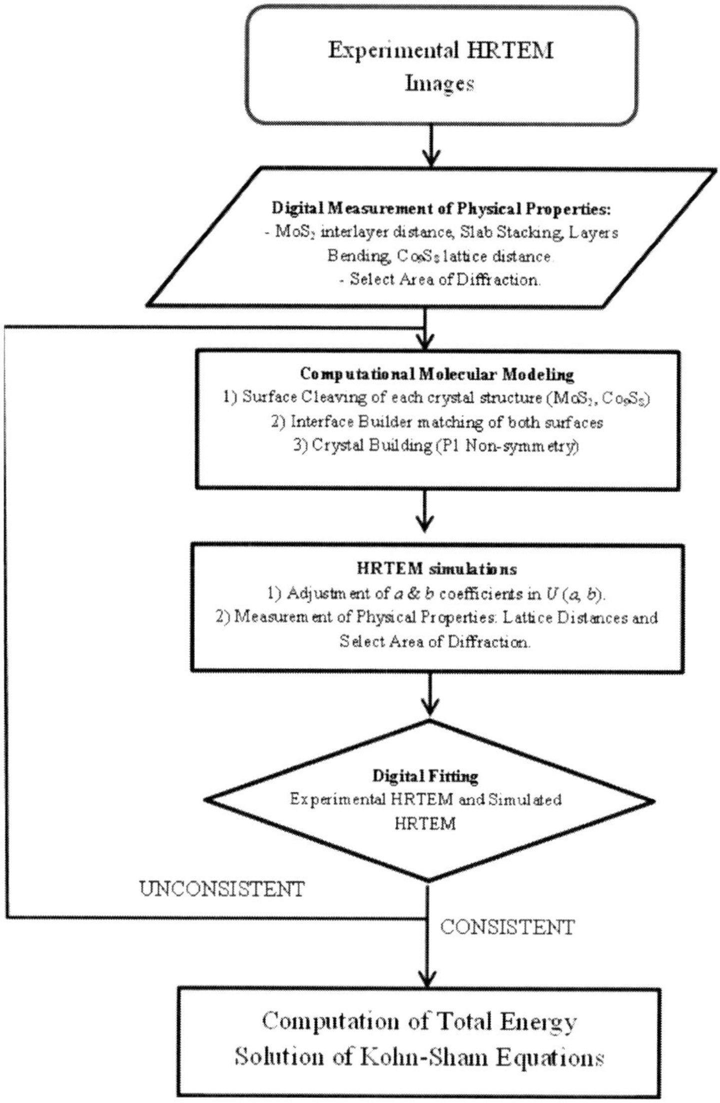

Figure 1: Chart flow of digitalization for HRTEM image reconstruction and molecular modeling used to produce this work.

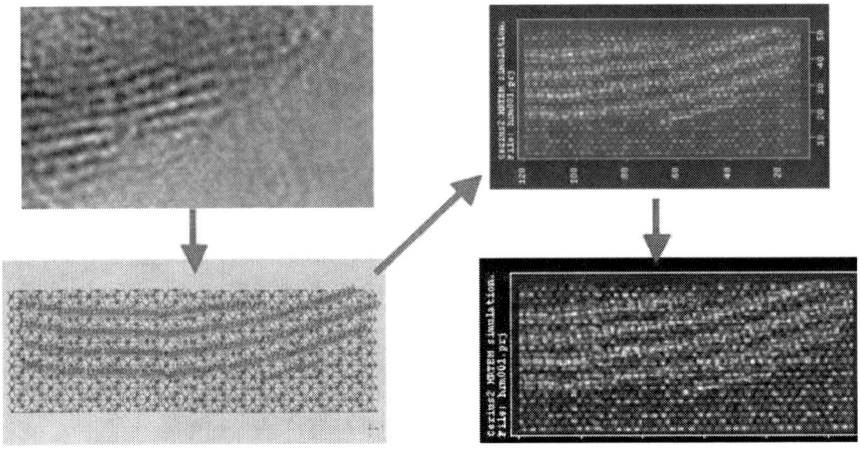

Figure 2: (a) Experimental HRTEM of (101)-MoS_2 stacked into (111)-Co_9S_8. (b) molecular model prepared for TEM simulation. (c) TEM simulation produced using projected potential with $\boldsymbol{a} = \boldsymbol{b} = 256$. (d) Digital superimposed to fit molecular model with simulated HRTEM.

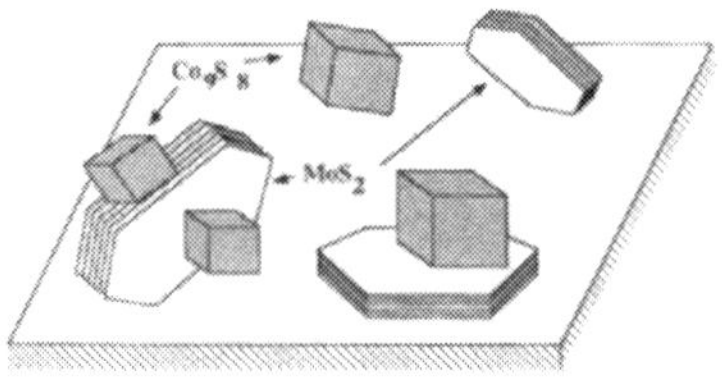

Figure 3: Schematic drawing of synergic contact model of Delmon [16].

All molecular modeling is to described black-white images taken with HRTEM instrument. A similar work is proposed in the past by Stockmann et al. using TEM simulations performed in MacTempas and Nelson using EELS tomography [18, 19], however, neither of both authors are able to localized the epitaxial contact between (101)-MoS_2 slabs and (111)-Co_9S_8 surface. The HRTEM simulations presented here are performed such that atoms are treated as frozen ionic cores [20], in consequence a less noisy and better quality of the image is produced, and relevant in understanding this promotion effect in transmission metal catalysis.

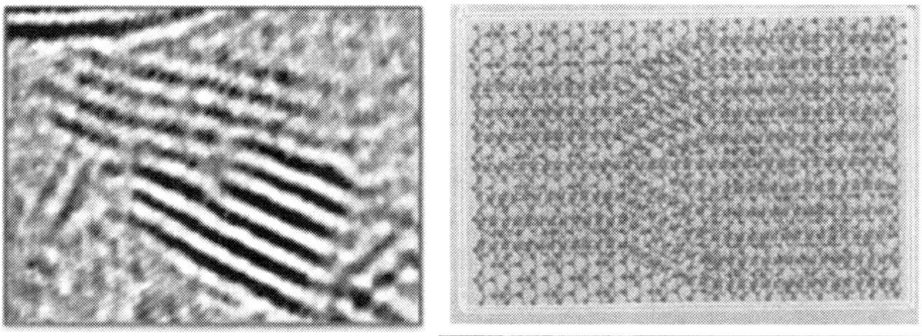

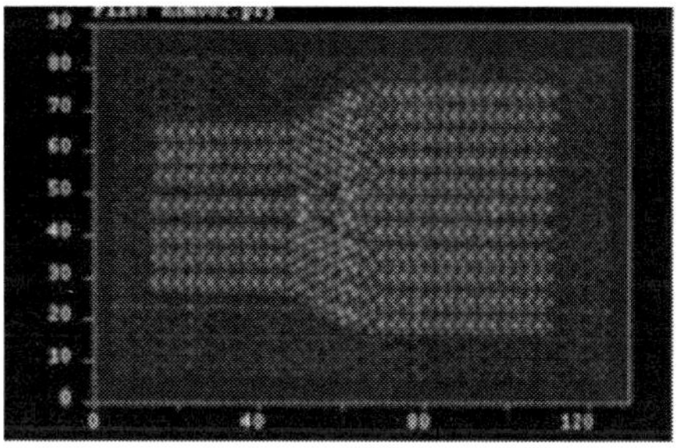

Figure 4: (a) Experimental HRTEM of MoS_2 indicating a dislocation of slabs. (b) Molecular model used to create TEM simulations. (c) Simulated HRTEM.

A recent previous study done by our group indicates the electronic character of this catalytic interface, results indicate a metallic character as described by Density of States and Band Structure using CASTEP code [21]. Now the question arises in two directions: (1) Localization of hydrogenation sites in this new catalytic interface or synergic contact model (figure 4) and (2) To find how tall the MoS_2 slabs are when they are attached to the (111)-Co_9S_8 structure.

CONCLUSIONS

Successful computational assisted HRTEM simulations are presented here, in order to resolve the layered MoS_2 catalytic structure. The observed experimental HRTEM indicates a contact between (101)-MoS_2 and (111)-Co_9S_8 planes in agreement with prediction by Delmon in 1988. Moreover, the chemical bonding among Mo and Co species, can lead to explain nature of promotional effect in bulk catalyst as described by DFT calculations [21]. With this combination between information gather from experimental and simulated TEM serves as best approach to resolve the intriguing question of promotion effect in catalytic layered transition metal sulfides. Finally, now the question arises in actual location of hydrogenation sites of DBT in HDS reaction, which can be approach using TS search of Dmol3 code.

ACKNOWLEDGMENTS

The authors want to thank: Programa de Mejora para el Profesorado (PROMEP) Mexico for economical support and IIT-UACJ. The Microelectronics Research Center of UT-Austin, Consejo Nacional de Ciencia y Tecnología-Mexico, Universidad Metropolitana and MRTI of

The University of Texas at El Paso.

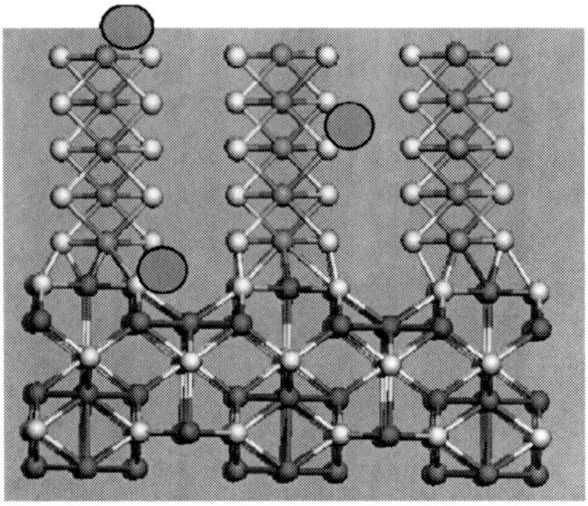

Figure 5: Side view of synergic contact between (101)-MoS_2 and (111) Co_9S_8, red dots represents possible selective sites for hydrogenation of DBT molecules in the HDS catalytic reaction.

REFERENCES

1.- G.C.A. Schuit et al., Applied Catalysis A: General 322.
2.- H. Topsøe et al., Springer, N.Y., (1996), p. 11.
3.- Chianelli, R. R. et al., Applied Catalysis, A, **227**, (2002)
4.- Applied Catalysis A: General 322 (2007).
5.- A. Vazquez, Materials Letters **35**, (1998).
6.- R. R. Chianelli et al., Catalysis Today, **23**, (1995).
7.- H. Topsøe, R. et al., Bull. Soc. Chim. Belg.93 (1984) 783
8.- Henrik Topsøe et al., Journal of Catalysis, Vol. 68, **2**, (1981).
9.- J.V. Lauritsen et al., Journal of Catalysis, Vol. 221, **2**, (2004).
10.- P. Raybaud et al., Surface Science, Vol. 407, **1-3**, pp. 237-250 (1998).
11.- Stockman, R.H. *et al.* Journal of Catalysis, 102, (1995).
12.- A. Gómez-Rodríguez et al., Ultramicroscopy, Vol. 110, **2**, (2010).
13.- Mohammad H. Siadati et al., Applied Catalysis A: General, Vol. 305, B, (2006).
14.- R. W. Phillips and A. A. Fote, *J. Catal.*, 1976, **41**, 168.
15.- R.R. Chianelli and G. Berhault, *Catal. Today*, 1999, **53**, 357.
16.- R.R. Chianelli, M. Daage, M.J. Ledoux, Advances in Catalysis, Vol. 40 (1994).
17.- Wells, A.F. Oxford: Clarendon Press, (1984).
18.- Stockmann et al. J. of Molecular Cat. A: Chemical 102, (1995).
19.- Florence Nelson et al., Microsc. Microanal. **16**, 194–199, (2010).
20.- J. P. Perdew, et al. Phys. Rev. B: Condens. Matter Phys., 1992, **46**, 6671.
21.- Ramos et al. RSC, Cat. Sci. & Tech. (In publication), (2011).

Mater. Res. Soc. Symp. Proc. Vol. 1373 © 2012 Materials Research Society
DOI: 10.1557/opl.2012.316

EFFECT OF HEAT TREATMENTS ON THE MICROSTRUCTURE AND PROPERTIES OF Al/SiC$_p$ COMPOSITES PREPARED BY NON-ASSISTED INFILTRATION

R. Martínez-López [(1)], M. I. Pech-Canul[(1)], Z. Chaudhury[(2)], L. A. González[(1)]

[1]Centro de investigación y de Estudios Avanzados del IPN Unidad Saltillo, Carretera Saltillo-Monterrey Km 13.5, Ramos Arizpe, Coahuila, México, C.P. 25900
[2] Materion Corporation, Advanced Materials Technologies and Services, 5520 Midway Park Pl Ne Albuquerque, NM 87108, USA.

ABSTRACT

Al/SiC$_p$ composites fabricated by the non-assisted infiltration route are attractive materials for various engineering applications. However, the presence of thermal stresses can impair their mechanical properties if they are utilized directly after processing. Therefore, heat treatments are potential solutions to this problem. In this work, the effect of T6-heat-treatment on the microstructure and hardness of Al/50% SiC$_p$ composites prepared by the non-assisted infiltration route is investigated. Previous to preform preparation, the SiC powders are coated with colloidal SiO_2. Infiltration tests are conducted using two experimental Al-Si-Mg alloys. The composites are sectioned in specimen sizes of 1 cm^2 and prepared using standard metallographic procedures. Then they are heat treated performing a solution treatment at 350°C for 3 h and artificial aging at 170°C for 1, 3 and 5 hours. In addition, the specimens are characterized by X-ray diffraction (XRD), optical microscopy, scanning electron microscopy (SEM) and energy dispersive X-ray spectroscopy (EDS). They are also characterized in hardness tests, comparing the behavior for the alloy with and without the various treatments. The results show that heat treatments do not affect the matrix/reinforcement interfacial condition and that the undesirable Al_4C_3 phase is not developed as consequence of thermal treatments. Hardness tests show that for alloy 1 the maximum hardness value is for the solution-heat-treated samples at 350°C and 5 h aging (28.26±4.02 HC); for alloy 2 the maximum hardness is achieved with solution heat treatment at 350°C and 3 h aging (25.86±4.05 HC). For composites processed with alloy1, the maximum hardness is obtained for the solution treated sample at 350 ° C for 3 h (91.8±2.17 HC), whereas for composites processed with alloy 2 the peak hardness is obtained in solution treated samples and aged at 170°C for 1 h (95.2±0.94 HC).

INTRODUCTION

Aluminum (and its alloys) is within the wide range of materials used as metal matrix in composite materials. It has low density, environmental resistance, good mechanical properties, in addition to its relatively high melting point that meets several engineering requirements. On the other hand, its melting point is low enough for processing so that these properties have made this kind of materials widely used in the automotive, aerospace, military, electronics and sports industries. Al MMCs (Metal Matrix Composites) reinforced with silicon carbide (SiC) particles are very attractive for structural and nonstructural applications at room temperature, as well as at relatively high temperatures [1, 2]. However, the massive commercialization of Al/SiC$_p$ (where p stands for particles) composites has not been widespread due to high production costs and technological barriers. There are several processing routes for fabrication of Al MMCs, including squeeze casting, spray deposition and infiltration. The latter can be performed in two modes, assisted and non-assisted, which differ in the way the liquid metal is incorporated into the porous SiC preforms. The former makes use of external pressure while the latter is based on capillary action. Although assisted infiltration is satisfactory in many respects, it has some disadvantages, such as the deformation or rupture of the SiC preform, in addition to the high costs of processing equipment. Thus, the dependence of non-assisted infiltration on capillary action makes it a simple and economic processing route. One variant of non-assisted infiltration is reactive infiltration, in which some reinforcements are formed in situ through the occurrence of reactions during infiltration [3]. It is understandable that the presence of thermal stresses developed during solidification in the processing, might impair the mechanical properties of composites produced by non-assisted infiltration. For example, phases with angular or acicular geometry are undesirable because they adversely affect the mechanical properties of the material. The solution and aging treatments have a beneficial effect on stress relief and modification of the morphology of certain phases observed in the microstructure after solidification. Accordingly, solution treatments and artificial aging, promise to be both economic and innovative processes for obtaining composites with better mechanical properties. In this work, the effect of solution treatment and artificial aging on the microstructure and mechanical properties of Al/SiC_p composites obtained by pressureless infiltration is studied.

EXPERIMENTAL DETAILS

Two Al-Si-Mg alloys fabricated and characterized (by spark emission and atomic absorption) are used in this investigation. The alloys are also characterized using differential thermal analysis (DTA) and thermogravimetric analysis (TG) to determine their melting points. The chemical compositions of the alloys are shown in Table I.

Table I. Chemical composition of the alloys used in the investigation (wt. %).

Alloy	Si	Fe	Cu	Mn	Mg	Al
Alloy 1	10.33	0.44	0.101	0.794	17.75	70.83
Alloy 2	21.71	0.385	0.006	0.012	15.30	62.58

Two types of SiC powders are used for the preparation of the porous preforms. While some powders are coated with colloidal silica, one batch of powders remains uncoated. The preforms are prepared by placing 26.8 g of SiC powders in a mortar and adding 4 wt. % dextrin as binder, followed by thoroughly mixing with about 5 ml of distilled water. The homogenous mixture is placed inside a rectangular-shaped metal mold and uniaxially compacted at a pressure of 9.5 MPa in a Carver model 4350L hydraulic press, to obtain preforms with 50% porosity with dimensions of 8cm x 3cm x 0.7 cm. Then the preforms are thermally treated in a Thermolyne 1400 furnace to remove the water and partially eliminate the dextrin by pyrolysis, first at 150 °C for 60 min and then at 300 °C for 60 min.

Infiltration tests are conducted in a horizontal Thermolyne 59300 model furnace equipped with an alumina tube with dimensions of 6.5cm in diameter and 76cm long. The SiC preforms are placed in molds with 50 g of the alloy and mold/metal/preform system is placed in the center of the tube furnace. After leak tests, the samples are heated at a rate of 15°C/min up to the test temperature (1200 °C) and maintained at that temperature for 120 min, conducting the infiltration test under nitrogen atmosphere at a flow rate of 10 cm^3/min. Then the system is cooled down to room temperature also in nitrogen at the rate of 15°C/min. After the infiltration tests, the molds with the composites are removed from the tube furnace for metallographic preparation and the microstructural characterization.

The composites are characterized by X-ray diffraction (XRD), optical microscopy (OM) and scanning electron microscopy (SEM) using a Phillips XL 30ESEM equipment coupled with an EDAX device to perform analysis by energy dispersive X-ray spectroscopy (EDS). The composites are prepared using standard metallographic procedures. First they are sectioned in 1 cm^2 specimens using a Struers Accutom-5 model, with diamond cutting wheels. Then they are ground using 120, 320, 500, 600, 800, 1200 and 2400 grit size sandpaper. Finally, the specimens are polished in nylon clothes using diamond pastes of 1, 0.5 and 0.25 μm.

RESULTS AND DISCUSSION

Figure 1 shows results from the analysis by DTA/TG of alloys 1 and 2, observing endothermic events around 580 °C and 620 °C, respectively. These events are attributed to the melting points of alloys. Furthermore exothermic events are recorded between 1050 °C and 1160 °C in alloy 1, and 1020 °C and 1110 °C in alloy 2. In both cases these events can be ascribed to the formation of Mg_2Si and AlN, respectively. The difference in formation temperatures of both phases in both alloys is due to content variation of Mg and Si.

Equations (2) to (4) show the results of thermodynamic calculations of reactions to form AlN and Mg_2Si phases at the temperature of 1200 °C.

$$2Al_{(l)} + N_{2(g)} = 2AlN \quad (1)$$

$$\Delta G_{1200°C} = -359.298 \text{ kJ/mol} \quad (2)$$

$$\Delta H = -658.334 \text{ kJ/mol} \quad (3)$$

$$Si_{(l)} + 2Mg_{(l)} = Mg_2Si \quad (4)$$

$$\Delta G_{1200°C} = -4678.645 \text{ kJ/mol} \quad (5)$$

$$\Delta H = -910.4469 \text{ kJ/mol} \quad (6)$$

Figure 2 shows X-ray diffraction patterns of composites made out from alloys 1 and 2. The phases present in the composite with alloy 1 are Si, AlN, Al, and SiC. The composite made from alloy 2 is characterized by the presence of the spinel ($MgAl_2O_4$) phase and magnesium silicide (Mg_2Si), due to the large amounts of Mg in the alloy, in addition to the previously mentioned phases. It should be noted that in both composites, the SiC particles are coated previously with colloidal SiO_2, which explains the absence of the undesirable aluminum carbide (Al_4C_3) phase. The Si available in both the alloy and the colloidal SiO_2 prevents formation of this phase.

Figure 3 shows the microstructure of two composites heat treated by a T6 treatment with 3 h of solution treatment. They correspond to composites without heat treatment (Fig. 3a), 3 h solution treatment at 350 °C and 1 h artificial aging at 170 °C (Fig. 3b), solution treatment during 3 hr solution treatment at 350 °C and 3hr of artificial aging at 170°C (Fig. 3c) and solution treatment during 3 h solution treatment at 350°C and 5 hr. of artificial aging at 170 °C (Fig. 3d). It is observed that there is no separation between the matrix and the reinforcement due to the heat treatments so that it can be concluded that there is a good interfacial condition between the aluminum matrix and SiC particles.

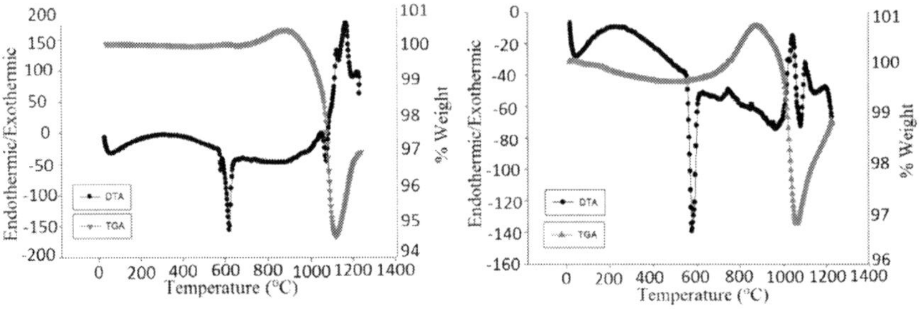

Figure 1. DTA/TGA thermograms of the two alloys used in the infiltration process using: a) alloy 1 and b) alloy 2.

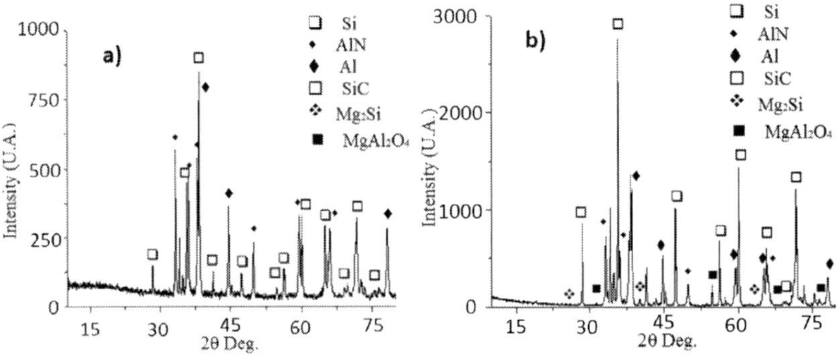

Figure 2. X-ray diffraction patterns of composites obtained using: a) alloy 1 and b) alloy 2.

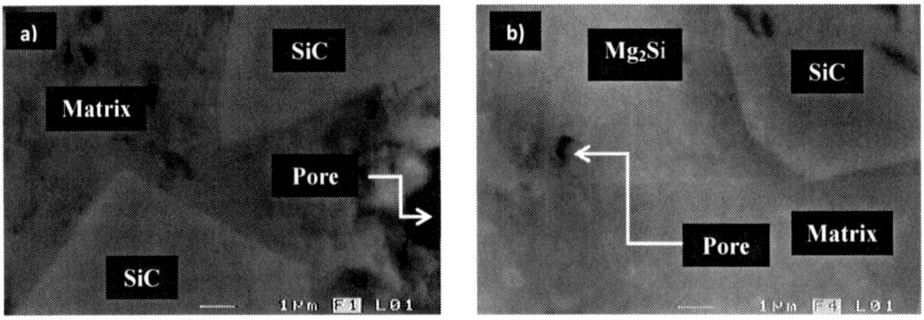

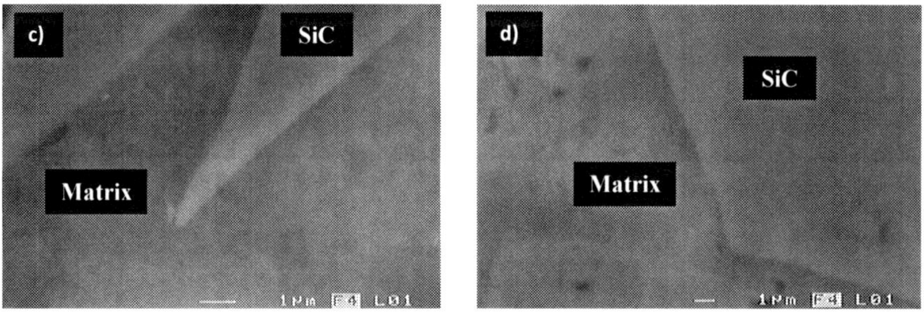

Figure 3. SEM images showing the microstructure evolution of Al/SiC$_p$ composites: a) without heat treatment, b), c) and d) subjected to 3 h solution treatment at 1, 3, and 5 h artificial aging, respectively.

Figure 4 shows results from surface hardness measurements of the alloys without reinforcement and the composites with solution treatment followed by artificial aging. It is evident that without the reinforcements and heat treatments, the alloys hardnesses are far below that of the composites. Moreover, the composite that is not heat treated (UTMMC) presents a lower hardness than the samples that are subjected to heat treatment (Aging 1h, aging 3h and aging 5h). On the other hand, the composite processed using alloy 2 shows that the highest surface hardness is obtained in the composite with solution treatment for 3 h at 350°C and aged for 1 h artificially at 170°C.

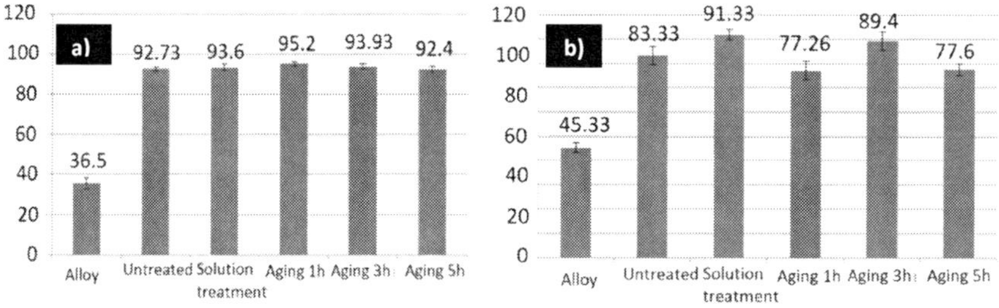

Figure 4. Superficial hardness (Rockwell B) of composites obtained using: a) alloy 1 and b) alloy 2.

CONCLUSIONS

• The XRD analysis shows that under the experimental conditions used, the phases formed in the Al/SiC composites during processing are AlN, Mg_2Si and $MgAl_2O_4$.

• The absence of the undesirable Al_4C_3 phase is explained by the beneficial effect of coating the SiC particles with colloidal SiO_2.

• Characterization by SEM shows that the heat treatments do not alter the reinforcement-matrix interfacial condition. Physically there is no separation between the matrix and reinforcement,

• Analysis of the effect of heat treatment on the surface hardness (Rockwell C) shows that solution treatment slightly increases the value of the hardness of composites obtained from alloy 1, while, the composites made out from alloy 2 which exhibited higher surface hardness are solution treated and aged for 1 h.

Acknowledgements

Mr. Martinez-López gratefully acknowledges Conacyt (Consejo Nacional de Ciencia y Tecnología) in Mexico for granting a scholarship. The authors also thank Mr. S. Rodriguez-Arias and Mr. M. A. Aguilar-Gonzalez for their assistance during the characterization by XRD and SEM, respectively.

References

[1] J.R. Davis. *"Aluminum and Aluminum Alloys"*. ASM Specialty Handbook, The Materials Information Society, Materials Park, OH, (1993) pp. 156-159.

[2] S. Suresh, A. Mortensen, A. Needleman. *"Fundamentals of Metal Matrix Composites, First Ed."*. Butterworth-Heinemann, Stoneham, MA, (1993) pp. 400.

[3] R. Escalera-Lozano. *"Estudio de la degradación de compósitos* $Al/SiC_p/MgAl_2O_4$ *procesados por infiltración reactiva a partir de aluminio de reciclaje, ceniza volante y ceniza de la cascarilla de arroz"*. Tesis de maestría, Centro de Investigación y Estudios Avanzados del IPN-Unidad Saltillo, 2005.

Mater. Res. Soc. Symp. Proc. Vol. 1373 © 2012 Materials Research Society
DOI: 10.1557/opl.2012.317

The Influence of Thermal Annealing on Microstructure and Mechanical Properties in High Performance Liquid Crystal Copolyesters

Adriana Reyes-Mayer [1,2], Amaury Constant [3], Angel Romo-Uribe[1,*] and Michael Jaffe [4]
[1]Laboratorio de Nanopolimeros y Coloides, Instituto de Ciencias Fisicas, Universidad Nacional Autonoma de Mexico, Cuernavaca, Mor. 62210, MEXICO.
[2] Facultad de Ciencias, Universidad Autonoma del Estado de México, Instituto Literario #100, Toluca, Edo de Mexico, C.P. 50000, MEXICO.
[3] Faculte des Sciences d'Orsay, Universite Paris Sud XI, Orsay, Cedex, FRANCE.
[4] New Jersey Institute of Technology, Newark NJ, U.S.A.
* Contact author: aromo-uribe@fis.unam.mx

ABSTRACT

In this research we have focused on the influence of thermal treatment for periods of time on the thermal and mechanical properties of extruded films of a series of high-performance thermotropic liquid crystal polymers (LCPs). The dependence of microstructure, thermal and mechanical properties on the extent of thermal treatment is investigated. Especially synthesized wholly aromatic LCPs based on hydroxybenzoic acid (B), hydroxynaphthoic acid (N), terephthalic acid (TA) and biphenol (BP) are kindly supplied by Hoechst Celanese Research Corp in the form of 50 μm thick extruded films. Thus, the influence of monomer composition is also studied in order to contrast the influence of molecular conformation. Thermal treatments are carried out at temperatures close to the solid-to-nematic transition ($T_{s\rightarrow n}$) for up to several hours under dry air conditions. The results show a profound influence of thermal annealing on morphology and mechanical modulus when annealing is carried out c.a. 40°C below $T_{s\rightarrow n}$, where solid-to-nematic transition and Young's modulus are significantly increased.

INTRODUCTION

Thermotropic liquid crystalline polymers (LCPs) have received considerable attention since the first report by scientists at Eastman Co. [1]. Since then the most successful LCPs, from the commercial point of view, are those developed by the Celanese Co. and commercialized by Ticona. The commercial LCPs display remarkable physical properties including tolerance for high temperatures, resistance to chemicals and acids, and high tensile strength [2,3]. These properties make them ideal for applications into the electronics, medical, dental, automotive and aerospace industries [4,6]. Thermotropic LCPs are wholly aromatics, i.e., there are not flexible spacers in the backbone (for instance, methylene CH_2 groups). Manipulating the orientation of the anisotropic macromolecules optimizes mechanical properties of LCP fibers [2]. Molecular alignment and microstructure are controlled through processing conditions during fiber spinning [7] and heat treatment [8,10]. At the molecular level the degree of straightness in the molecular architecture of the polymers studied here (figure 1) depends upon the amount of 2,6-hydroxynaphthoic acid (N), which upsets the straight conformation of the chain. The double aromatic ring in the N monomer acts as a parallel offset thereby disturbing the regularity of molecular packing, thus lowering the solid-to-nematic transition temperature [2]. Despite the evident importance and striking physical properties of commercial thermotropic LCPs, there are to date few studies addressing the influence of thermal treatment on their mechanical properties

[7,9]. In this work we have investigated the influence of thermal treatment over extended periods of time on the thermal and mechanical properties thermotropic copolyesters with varying concentration of HNA. The mechanical properties of the reinforced LCPs are studied under uniaxial tension mode and the results are correlated with the thermal properties.

EXPERIMENT

Materials. Extruded thin films of three thermotropic liquid crystalline copolyesters are kindly supplied by the former Hoechst Celanese Research Corporate. These are termed B-N, COTBP and RD1000. B-N is a copolyester based on random units of ~73 mol % B and ~27 mol % N. The chemical formula is shown in Figure 1a. This material is marketed by Ticona under the tradename of *Vectra A*®. COTBP (Figure 1b) is a random copolyester made of 66 mol% 1,4-hydroxybenzoic acid (B), 5 mol%2,6-hydroxynaphthoic acid (N), and 17.5 mol% (equal parts) of terephthalic acid (TA) and biphenol (BP). Finally, RD1000 is a wholly aromatic thermotropic LCP whose chemical composition is very similar to COTBP, however its specific chemical composition is proprietary information of Celanese Co.

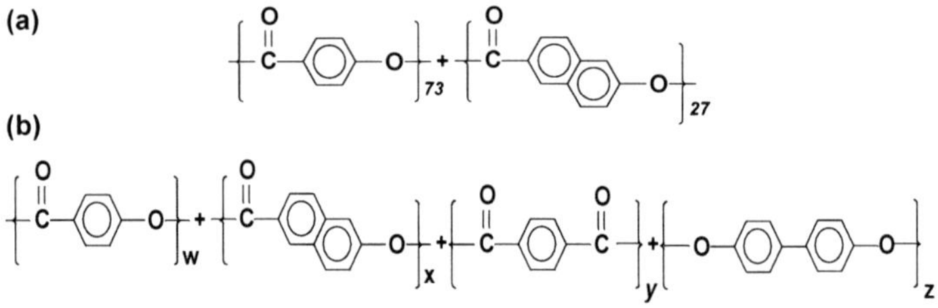

Figure 1. Chemical formulas of thermotropic LCPs. (a) B-N, (b) COTBP.

Thermal analysis. The thermal transitions are determined by differential scanning calorimetry (DSC). For DSC experiments the DSC6000™ calorimeter manufactured by Perkin Elmer (Connecticut, USA) is used. Temperature and enthalpy calibration are carried out using analytical grade indium (T_m = 156.6 °C) and corrections are made for the instrument 'baseline'. The thermal transitions are determined at a heating rate of 20 °C/min.

Microstructure. Polarized optical micrographs are obtained using a Leitz Laborlux D optical microscope. Small-angle light-scattering patterns in H_V polarization condition are obtained using an in-house instrument and a He-Ne laser.

Mechanical properties. The mechanical properties are studied under uniaxial tension. Tests are carried out at room temperature on a universal testing machine Instron 4206 (USA) with a crosshead speed of 5 mm/min.

SEM. The evaluation of the microfillers adherence as well as the fracture mode in the fractured surfaces is carried out utilizing scanning electron microscopy (SEM). A SEM model JEOL JSM-5900LV is used and the sample's fracture surfaces are metal coated with Au/Pd by a sputtering procedure.

DISCUSSION

Figure 2 shows photographs of the as-received extruded films of the LCPs. B-N is optically opaque due to the high density of defects (disclinations). On the other hand, COTBP and RD1000 are nearly transparent, denoting a uniaxially oriented defect texture, as confirmed by POM and SALS.

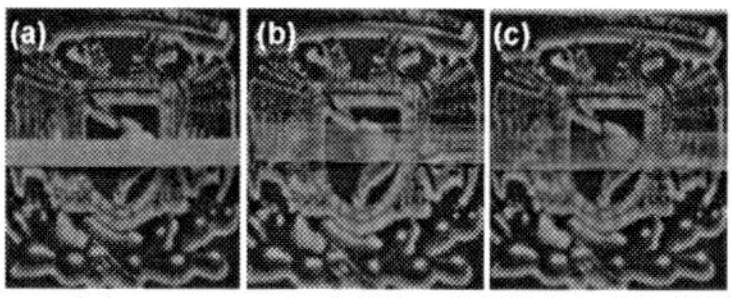

Figure 2. Extruded films of thermotropic LCPs. (a) B-N, (b) COTBP, and (c) RD1000.

Figure 3 shows polarized optical micrographs and a SALS pattern of the as-received extruded film of COTBP, exhibiting a highly oriented texture. Notice the line defects (dark horizontal lines in Figure 2b) oriented along the extrusion direction (horizontal, as indicated by the arrow in Figure 1a). The SALS pattern is anisotropic, with an elongated intensity on the axis orthogonal to the extrusion direction. The anisotropy denotes the degree of orientation of the defects and the azimuthal spread denotes the degree of alignment of these defects.

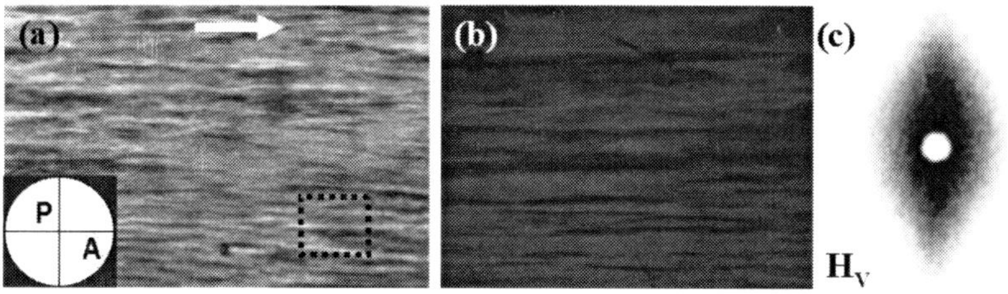

Figure 3. (a) Polarized optical micrograph of extruded thin film of COTBP. (b) POM micrograph, from dotted region in (a). (c) Corresponding SALS pattern, utilizing H_V polarization.

Figure 4a shows POM micrographs of RD1000 as-received film, it shows a well-defined line structure oriented along the extrusion direction (indicated by the arrow). This oriented texture arises from the mesophase defects having been oriented under the shear flow experienced in the extrusion die. The corresponding SALS pattern is anisotropic with intensity elongated on the axis orthogonal to the extrusion. The azimuthal spread of intensity of the SALS patterns show that the texture is better oriented in RD1000 than in COTBP films.

On the other hand, the extruded film from B-N is optically opaque and yellowish, as shown in Figure 1a, and therefore it is not possible to carry out POM and SALS studies.

Heat treatment is applied to all samples for a period of several hours under air conditions. B-N is treated at 240°C, COTBP is annealed at 300°C and RD1000 is annealed at 300°C, that is, at $\Delta T=T_a-T_{s\rightarrow n}$=40°C. Figure 5 shows typical DSC first heating scans obtained for B-N. The as-received film shows a broad and weak endotherm at $T_{s\rightarrow n}$=280°C. However, it can be seen that during the first hour of annealing a second endotherm appears at ca. 262°C. The $T_{s\rightarrow n}$ weakens significantly, but remains at around 280°C. Interestingly, after 3 hours annealing of B-N exhibits two endotherms. The low temperature endotherm consistently shifts to lower temperatures whereas $T_{s\rightarrow n}$= remains constant. This thermal behavior has been previously reported for B-N [11].

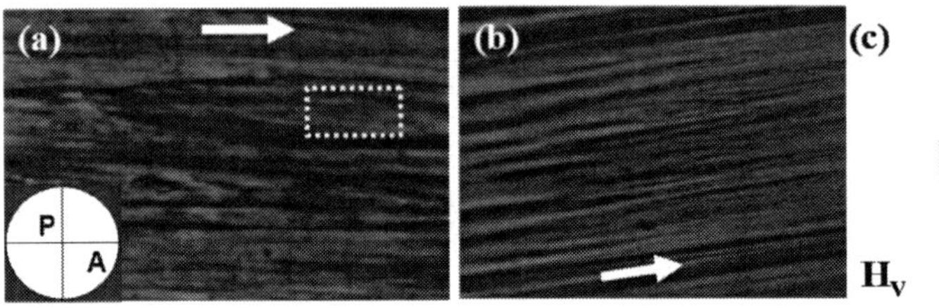

Figure 4. (a) Polarized optical micrograph of extruded thin film of RD1000. (b) POM micrograph, from dotted region in (a). (c) Corresponding SALS pattern, utilizing H_V polarization.

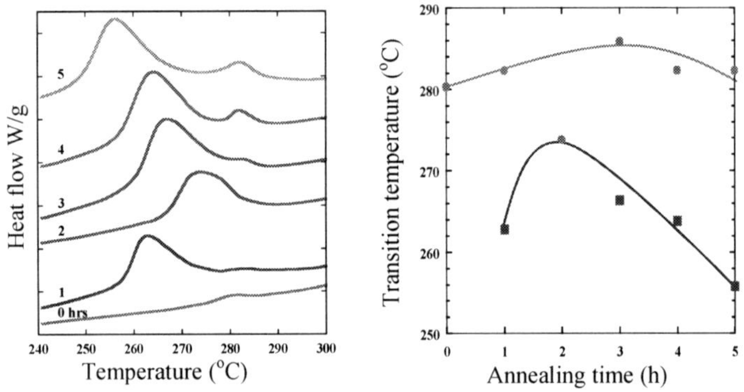

Figure 5. B-N thermotropic LCP, extruded film. (a) DSC first heating scans of films heat-treated at 240°C. (b) Thermal transition as a function of annealing time.

Uniaxial tensile testing is carried out, stretching at 5 mm/min at room temperature, on as-received and 3 hours heat treated films. The mechanical behavior of the LCP films is characterized by a brittle behavior, with a steep linear elastic region. Figure 6a shows a plot of Young's modulus. The results show significant increase of mechanical modulus after the extruded films are thermally annealed. Moreover, the mechanical tests show that the fracture of the films always occurs within 3% strain and is characterized by fibrillation of the films, as shown by the SEM micrographs in Figure 6b.

CONCLUSIONS

The results show that transition temperatures of the liquid crystal copolyesters are significantly influenced by thermal annealing. There is an increase of solid-to-nematic transition ($T_{s\rightarrow n}$) or the production of a second endotherm, when annealing is carried out at $\Delta T = T_a - T_{s\rightarrow n}$=40°C, T_a being the annealing temperature. Young modulus exhibits a significant increase leveling off after the first three hours of thermal treatment. Thermal treatment also produces a change in fracture mode, from ductile to brittle. The brittle mode displays a fiber-like morphology, as revealed by SEM.

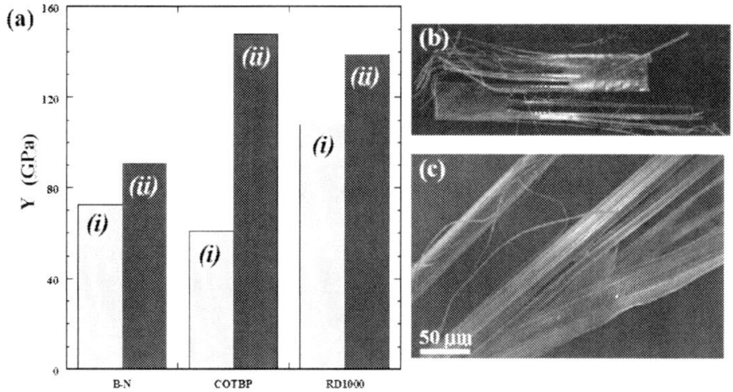

Figure 6. (a) Young's modulus of (i) as-received, and (ii) heat-treated thermotropic LCPs. Samples are treated for 3 hrs. (b) Picture of RD1000. (c) SEM micrograph of RD1000, taken from the dotted region marked in (b).

ACKNOWLEDGMENTS

A. Reyes-Mayer acknowledges the financial support of the COMECYT scholarship program. Thanks to Hoechst Celanese Corp (Summit, NJ, USA) for providing extruded films. This research is supported by CONACYT, under the program CIAM2008, grant 107294.

REFERENCES

1. W.J. Jackson and H. F. Kuhfuss, J. Polym. Sci. Polym. Chem. 14, 2043 (1976).
2. G.W. Calundann and M. Jaffe, Synthetic polymers 15-17, 247 (1982).
3. H.N. Yoon, L.F. Charbonneau, G.W. Calundann, Adv. Mat. 4, 206 (1992).
4. L. Czyborra and F. Metzmann, Kunststoffe 88, 721 (1998).
5. L.C. Sawyer, H.C. Linstid and M. Romer, Plastics Engineering 54, 37 (1998)
6. A.A. Collyer, Elseiver Appl. Sci., (1992).
7. M. Cakmak, A. Teitge, H.G. Zachmann and J.L. White, J. Polym. Sci. Poly. Phys. 31, 371 (1993).
8. J.B. Stamatoff, Mol. Cryst. Liq. Cryst, 110, 75 (1984).
9. T.M. Nicholson and I.M. Ward, Polymer 39, 315 (1998).
10. A.M. Donald and A.H. Windle, Liquid Crystalline Polymers. Cambridge: Cambridge University Press. (1992)
11. A. Romo-Uribe, T.J. Lemmon and A.H. Windle, J. Rheol. 41, 1117 (1997).

Mater. Res. Soc. Symp. Proc. Vol. 1373 © 2012 Materials Research Society
DOI: 10.1557/opl.2012.319

Preparation of amorphous aluminium ammonium phosphates

Barbara Grzmil, Kinga Łuczka and Marta Gleń
West Pomeranian University of Technology, Szczecin, Institute of Chemical and Environment Engineering, 70-322 Szczecin, Pułaskiego 10, E-mail: barbara.grzmil@zut.edu.pl

ABSTRACT

A method for obtaining of amorphous aluminium ammonium phosphates is developed. The statistical program STATISTICA 9 is used for planning and evaluation of the experiments. Research is carried out according to the Box-Behnken model for the following values of independent variables (input factors): pH: 6.0 ± 2; concentration of reagents: 40 ± 10 wt%, molar ratio of NH_4^+:PO_4^{3-} in substrates (2 ± 1:1) i.e. Al^{3+}:NH_4^+:PO_4^{3-} (0.66:1-3:1). On the basis of this research, process parameters are determined, in which materials with the expected physicochemical properties (chemical composition, specific surface area, oil absorption number) can be obtained.

INTRODUCTION

The most commonly used pigments having good anticorrosive properties are minium (mixture of lead oxides) and zinc chromate. Due to the environmental reasons these pigments are no longer desired. Other pigments providing coatings with suitable anticorrosive properties are borates, molybdates, ferrites, zinc powder, and phosphates (I, II and III generation) [1,2].

It is claimed that the introduction of the molybdenum anion brings a beneficial effect to the increase of effectiveness of the phosphate pigments [3,4]. The anticorrosive effectiveness of phosphate pigments is tested in various types of paints [5-11]. Zubielewicz and Gnot [6] conclude that calcium zinc phosphate and zinc ferrite have the best performance as a result of steel surface passivation. It is shown [8] that generation II phosphates (aluminium zinc phosphate, iron (II) zinc phosphate, zinc phosphormolybdate), used in acidic medium, are characterized by the improved corrosion inhibiting properties which are similar to zinc chromate.

Aluminum phosphate is also used as an anticorrosive pigment. It is also used as a catalyst, catalyst support, high-temperature binder material, raw material for ceramics and optical glasses production or as an absorbent in medicine. Burrell et al. [11] pointed out the difficulty in obtaining of aluminum phosphates with repeatable properties. Mesoporous material with high surface area, narrow pore size distribution and high thermal stability is obtained [12]. In other publications [13-15], studies on the obtaining of amorphous, crystalline and nanocrystalline aluminum phosphate pigments are presented. Preparation and structure of aluminium ammonium dihydrate hydroxyphosphates used in molecular sieves are discussed [16]. Stability of aluminum phosphates in the NH_3-Al_2O_3-P_2O_5-H_2O system is studied [17]. In the publication [18], methods for obtaining of aluminium ammonium phosphates $(NH_4)_{2-x}(H_3O)_x\,Al_2(PO_4)_2(OH)_2$, $(NH_4)_3Al_2(PO_4)_3$ and $(NH_4)_3Al(PO_4)_2$ are described.

Anticorrosive properties of phosphates can result from their strong complexing power [5]. They can form complexes with both metal cations and functional groups of filmogen. Polyphosphates have better complexing properties than orthophosphates. Anticorrosive properties of Al ortho-, poly- and metaphosphates, Al Zn and Al Sr polyphosphates, or Al Ca, Ca Zn and Al silicapolyphosphates are tested in acrylic and styrene-acrylic coatings [19,20]. Naderi and Attar [21] have shown better protective properties of aluminium zinc polyphosphate in comparison with the hydrated zinc phosphate in their electrochemical studies.

The aim of the presented research is to elaborate a processing route for the preparation of aluminium ammonium phosphates which can be used as pigments in anticorrosive coatings. Therefore, physicochemical properties of obtained materials like chemical composition, oil absorption number and specific surface area are determined.

EXPERIMENT

Studies on the synthesis of aluminium ammonium phosphates are performed on the basis of the experimental planning and analysis according to the Box –Behnken plan for 15 experiments [22]. Process independent variables are: x_1 - pH of the reaction mixture (6 ± 2) x_2 - concentration of the reactants (40 ± 10 wt%) and x_3 - molar ratio of NH_4^+:PO_4^{3-} in substrates (2 ± 1:1) i.e. Al^{3+}:NH_4^+:PO_4^{3-} (0.66:1–3:1). Multiple dependent variables defining physicochemical properties of the obtained products are: Al/NH_3 molar ratio (y_2), oil absorption number (y_3) and specific surface area S_{BET} (y_4). Dependent variable y_1 is calculated on the basis of the molar ratio of Al^{3+}:NH_4^+:PO_4^{3-} in obtained phosphates (on the assumption that it is referred to 1 mole of PO_4^{3}) according to the following equation:

$$y_1 = \frac{3n_{Al^{3+}} + n_{NH_4^+}}{3} \tag{1}$$

where: n_{Al} - number of aluminium ion moles, $n_{NH_4^+}$ - number of ammonium ion moles per 1 mole of PO_4^{3-}.

Table 1. Three-level $3^{(k-p)}$ fractional factorial design.

no.	Independent variables			The molar ratio of products Al^{3+}:NH_4^+:PO_4^3:OH^-	Dependent variables			
	x_1	x_2	x_3		y_1	y_2	y_3	y_4
1	4	30	2	1.00 : 0.14 : 1 : 0.14	1.05	7.14	101	47
2	8	30	2	1.15 : 0.73 : 1 : 1.18	1.39	1.58	99	294
3	4	50	2	0.85 : 0.37 : 0.08 : 1*	0.97	2.30	80	30
4	8	50	2	1.47 : 0.94 : 1 : 2.35	1.78	1.56	98	101
5	4	40	1	1.95 : 0.79 : 1 : 3.64	2.21	2.47	82	127
6	8	40	1	1.27 : 0.51 : 1 : 1.32	1.44	2.49	82	241
7	4	40	3	1.35 : 0.32 : 1 : 1.37	1.46	4.22	89	64
8	8	40	3	2.20 : 1.36 : 1 : 4.96	2.65	1.62	94	335
9	6	30	1	1.07 : 0.55 : 1 : 0.76	1.25	1.95	96	35
10	6	50	1	0.86 : 0.14 : 0.28 : 1*	0.91	6.14	101	118
11	6	30	3	1.50 : 0.90 : 1 : 2.4	1.80	1.67	100	68
12	6	50	3	1.36 : 0.71 : 1 : 1.79	1.60	1.92	90	129
13	6	40	2	1.75 : 0.70 : 1 : 2.95	1.98	2.50	99	116
14	6	40	2	1.68 : 0.73 : 1 : 3.00	1.92	2.30	96	111
15	6	40	2	1.78 : 0.76 : 1 : 3.25	2.03	2.34	93	115

* Molar ratio Al^{3+}:NH_4^+:H^+:PO_4^{3-}

The aim of the assumed Box – Behnken plan is to identify significant factors influencing studied parameters and find suitable input values which allow obtaining a product with the expected properties. Factorial design and the results of experiments are presented in Table 1.

Reagent grade substrates $NH_4H_2PO_4$, $(NH_4)_2HPO_4$, $(NH_4)_3PO_4 \cdot 3H_2O$, $Al_2(SO_4)_3 \cdot 18H_2O$, H_2SO_4 (96 wt%) and aqueous ammonia (25 wt% NH_3) are used. The process for obtaining of aluminium ammonium phosphates is performed in a glass reactor at constant temperature of 23 °C. To the solution of ammonium phosphates, mixed with constant rate, a solution of aluminium sulphate is added. The reaction time is constant and is 30 min. It is followed by the separation of precipitate from the mother liquor and triple ishing with water (weight ratio of the liquid to solid phase is 3: 1). The obtained product is dried at 70°C for 3 hours.

Aluminium, phosphates' and ammonia contents are determined in the obtained products using complexometric titration, spectrophotometric method [23] and ion selective electrode Orion 11-35 type [24], respectively. Oil absorption number (grams of oil needed to produce a homogeneous paste with 100 grams of dry pigment being tested) is determined according to PN-EN ISO 787-5 [25]. In order to determine the phase composition of products X-ray diffraction analysis (X'Pert PRO Philips diffractometer, Cuk_α radiation) is used. The specific surface area S_{BET} is determined using Micrometrics Quadrasorb SI Quantachrome Instrument. N_2 adsorption/desorption measurements are carried out at liquid N_2 temperature.

RESULTS

Analyzing the phase composition of phosphates, obtained by X-ray diffraction analysis, it is found that, regardless of process conditions used, the amorphous solids are observed. There is only one broad 100% reflection peak at $2\theta=29.6$.

Investigations are conducted for three levels of factors which statistical analysis predestine the approximation of the operational variables impact on the basis of the linear (L) and quadratic (Q) models. It increases the evaluation reliability and allows estimating the interaction between the operating parameters [22,26].

Basing on the statistical analysis, which determines the process parameters having significant influence on the properties of obtained phosphates, the results (changes in the output values $y_{1,2,3,4}$ at extreme values of independent variables $x_{1,2,3}$) are evaluated. The evaluation of results, for the probability test, is performed at the significance level equal to 5%.

In order to demonstrate the presence of H^+ or OH^- ions in the obtained phosphates (Eq. 1) the ratio of mole sum of Al^{3+} and NH_4^+ cations is calculated. Valence of these ions to the valence of one PO_4^{3-} (y_1) mole is taken into account. In neutral aluminium ammonium phosphate this ratio is equal to 1. When the ratio is less than 1 hydrogenphosphates appear and when is greater than 1 hydroxyphosphates occur. The analysis of the y_1 value in relation to the synthesized phosphates showed that only products obtained in experiment 3 and 10 (Table 1) are characterized by the value of the dependent variable $y_1 <1$, which indicates most likely that they are aluminium ammonium hydrogenphosphates. In other cases (13 experiments) they are aluminium ammonium hydroxyphosphates with molar ratio of PO_4^{3-} to OH^- changing from 0.2 to 7.14. Therefore, in the rest of work aluminium ammonium hydroxyphosphates will be taken into consideration. Products in which the degree of substitution of phosphate groups with hydroxyl groups is the highest (lowest molar ratio of PO_4^{3-} ions to $OH^- = 0.20$) are obtained in experiment 8 (Table 1). The pH of the reaction mixture is 8.0, salt concentration is 40 wt% and the molar ratio of $Al^{3+}:NH_4^+:PO_4^{3-}$ in substrates is 0.66:3:1.

It is concluded that the value of the dependent variable (y_1), which reflects the type of aluminium ammonium phosphates (hydrogenphosphates, hydroxyphosphates), is the most affected, successively, by the pH of the reaction mixture ((x_1(L), effect 1.54; x_1(Q), effect 0.39 $p<0.05$), concentration of solution ((x_2(L), effect -0.057; x_2(Q), effect -0.616; $p<0.05$) and the molar ratio of reactants: $Al^{3+}:NH_4^+:PO_4^{3-}$ ((x_3(L), effect 0.43; x_3(Q), effect -0.026 $p<0.05$).

Analysis of the effect of process factors on the value of the molar ratio Al/NH_3 (y_2) in aluminium ammonium hydroxyphosphates shows a statistically significant interaction of only the pH of the reaction mixture (x_1). If the effect's sign is negative (pH (L), effect -2.22, $p<0.05$), the decrease of x_1 value will increase the value of output factor y_2. The molar ratio of Al/NH_3 in the obtained products ranged from 1.56 to 7.14. Aluminium ammonium hydroxyphosphates with the favourably highest ammonia contents (the smallest Al/NH_3 molar ratio) are precipitated from the reaction mixture with the pH=8.0 while the salt concentration is 50 wt% and the molar ratio of $Al^{3+}:NH_4^+:PO_4^{3-}$ in substrates is 0.66:2:1 (Table 1, experiment 4).

The influence of independent variables on the oil absorption number (y_3) of synthesized aluminium ammonium hydroxyphosphates is evaluated. It is found that the increasing pH value ((x_1 (L), effect 5.25, x_1 (Q), effect 7.25, $p<0.05$) can cause the increase of the output y_3 value. Oil absorption number of the obtained products is in the range 82 - 101 (Table 1, experiments 5 and 10) and it characterizes sorption properties of pigment. Moreover, it depends on the surface area and particle size distribution.

Output factor y_4 (specific surface area S_{BET}) is also evaluated. A significant influence of the pH (x_1 (L), effect 175.75, $p<0.05$) on the specific surface area is observed. It can be concluded that the increasing pH of the precipitation can cause the increase of the specific surface area (y_4) of the obtained aluminium ammonium hydroxyphosphates. The specific surface area of the obtained products varied from 30 to 335 m^2/g. The highest S_{BET} value is observed for aluminium ammonium hydroxyphosphates obtained in the experiment in which the pH of the reaction mixture is 8.0, salt concentration is 40 wt% and the molar ratio of $Al^{3+}:NH_4^+:PO_4^{3-}$ in substrates is 0.66:3:1 (Table 1, experiment 8).

In the statistical evaluation of the tested model, concentration and substrate molar ratio had no significant effect on the dependent variables $y_{1,2,3,4}$. However these factors can modify the strength of the effect of the pH on the Al/NH_3 molar ratio in the obtained hydroxyphosphates and their oil absorption number or specific surface area.

In order to confirm the significance of the evaluation results Anova variance analysis is performed. It is concluded that the most significant effect on the dependent variables $y_{1,2,3,4}$ had the pH. In Fig. 1, in order to identify and evaluate the significance of analyzed problems, Pareto charts of standardized effects, which illustrate the impact of the studied factors on the dependent variables y_1, y_2, y_3, y_4, are shown.

CONCLUSIONS

On the basis of the performed experiments, which concern the synthesis of aluminium ammonium hydroxyphosphates with the expected properties, it is assumed that the type of phosphates (hydrogenphosphates, hydroxyphosphates) and the degree of substitution of PO_4^{3-} with OH^- groups are significantly affected by the pH of the reaction mixture.

Moreover it is stated that the pH had the only significant effect on the Al/NH_3 molar ratio in phosphates, oil absorption number and specific surface area. It is confirmed that the pH should had the highest value in the tested range. This is important as long as aluminium ammonium hydroxyphosphates with high ammonia contents and high specific surface area are obtained. In the statistical evaluation of the tested model, concentration and substrate molar ratio had no significant effect on the dependent variables $y_{1,2,3,4}$. However these factors can modify the strength of the effect of the pH on the studied output factors in the obtained materials.

Statistically performed evaluation of the planned research simplifies the selection of process parameters in order to obtain aluminium ammonium hydroxyphosphates with the expected $y_{1\text{-}4}$ values of output variables, that is, projected physicochemical properties.

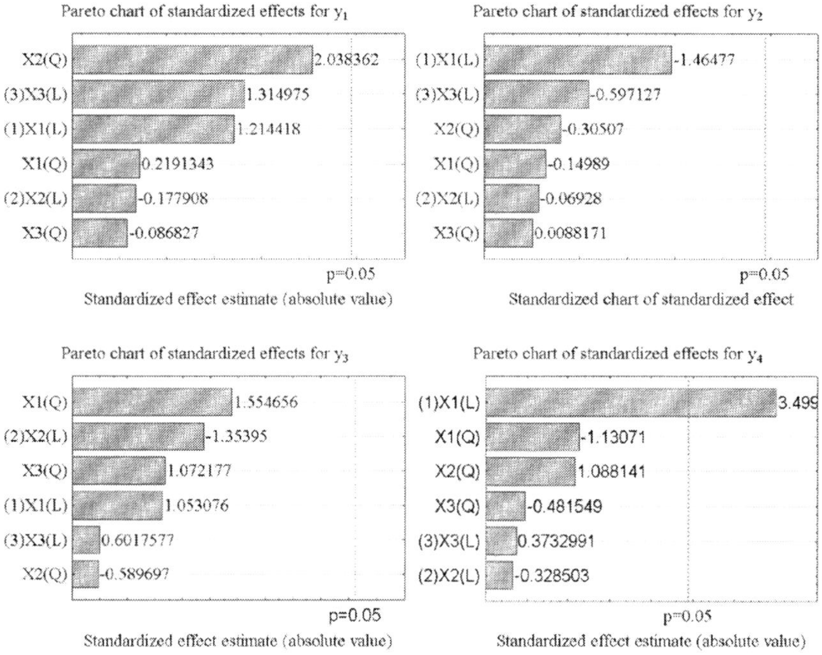

Figure 1. Pareto charts of standardized effects for y_1, y_2, y_3, y_4.

ACKNOWLEDGMENTS

This work is funded from financial support on science (POLAND) for 2011-2014

REFERENCES

1. T. Spychaj and S. Spychaj, *Farby i kleje wodorozcieńczalne*, WNT (Warszawa, 1996).
2. E. Śmieszek and E. Kamińska, *Ochrona przed korozją* **4**, 85 (1996).
3. B. del Amo, R. Romagnoli and V. F. Vetere, *J. Appl. Electrochem.* **29**, 1401 (1999).
4. L. Veleva, J. Chin and B. del Amo, *Prog. Org. Coat.* **36**, 211 (1999).

5. P. Mošner, A. Kalendova and L. Koudelka, *Dyes Pigm.* **45**, 29 (2000).
6. M. Zubielewicz and W. Gnot, *Prog. Org. Coat.* **49**, 358 (2004).
7. G. Blustein, M. C. Deya and R. Romagnoli, *Appl. Surf. Sci.* **252**, 1386 (2005).
8. M. Bethencourt, F. J. Botana, M. Marcos, R. M. Osuna and J. M. Sánchez, *Prog. Org. Coat.* **46**, 280 (2003).
9. P. de Lima-Neto and D. de Araujo, *Prog. Org. Coat.* **62**, 344 (2008).
10. M. A. Abd El-Ghaffar, E. A. M. Youssef and N. M. Ahmed, *Pigm. Resin Technol.* **33**, 226 (2004).
11. L. S. Burrell, C. T. Johnston, D. Schulze, J. Klein, J. L. White and S. L. Hem, *Vaccine* **19**, 275 (2001).
12. G. Liu, M. Jia, Z. Zhou, L. Wang, W. Zhang and D. Jiang, *J. Colloid Interf. Sci.* **302**, 278 (2006).
13. M. M. Beppu, E. C. de Oliveira and F. Galembeck, *J. Colloid Interf. Sci.* **178**, 93 (1996).
14. F. Lagno and G. P. Demopoulos, *Ind. Eng. Chem. Res.* **44**, 8033 (2005).
15. B. Kic, B. Grzmil and K. Lubkowski, *Przem. Chem.* **88**, 468 (2009).
16. J. B. Parise, *Acta Cryst.* **40**, 1641 (1984).
17. A. Ono and M. Kabayashi, *J. Mater. Sci. Letters* **6**, 178 (1987).
18. M. E. Medina, M. Iglesias, E. Gutièrrez-Puebla and M. A. Monge, *J. Mater. Chem.* **14**, 845 (2004).
19. *Pat.* EP 1878700 (2008).
20. A. Kalendova, *Anti-Corros. Method Mater.* **50**, 82 (2003).
21. R. Naderi and M. M. Attar, *Electochimica Acta* **53**, 5692 (2008).
22. G. E. P. Box, J. S. Hunter and W. G Hunter, *Statistics for experimenters*, Wiley - Interscience (Hoboken 2005).
23. J. Minczewski and Z. Marczenko, *Chemia analityczna,* PWN (Warszawa 2005).
24. Orion Ammonia Electrode, *Instruction manual*, Thermo Electron Corporation (2003).
25. Standard PN-EN ISO 787-5, 12.1999.
26. J. Pietraszek, *Elektroniczny Podręcznik Statystyki* (Kraków 2004).

Mater. Res. Soc. Symp. Proc. Vol. 1373 © 2012 Materials Research Society
DOI: 10.1557/opl.2012.320

Effect of manganese on the thermostability and reducibility of cobalt nanomaterials.

Zofia Lendzion-Bieluń
West Pomeranian University of Technology, Szczecin, Institute of Chemical and Environment Engineering, Pułaskiego 10, 70-322 Szczecin, Poland

ABSTRACT

Cobalt nanomaterials with promoters have been prepared by precipitation followed by calcination and impregnation method. The obtained materials are characterized by ICP, H_2-TPR, BET and XRD. A small addition of manganese increases the specific surface area of cobalt nanomaterials and thermostability under reduction atmosphere. Bulk Co_3O_4 has been reduced in two steps ($Co^{3+} \rightarrow Co^{2+} \rightarrow Co$). A small addition of manganese to the cobalt oxide shifts the reduction peaks to high temperatures.

INTRODUCTION

Nanocrystalline cobalt oxides and cobalt have been important subject for studies. These materials are applied in various fields of modern technology, including catalysts for various processes [1,2]. Catalytic properties of these materials depend on many factors e.g. preparation methods, promoters, structure and morphology. Specific surface area is a key factor to determine the activity of cobalt catalysts. Metallic cobalt is an active form of cobalt catalyst in more important industrial processes e.g. Fischer-Tropsch synthesis and methane stream reforming. A small addition of structural promoters, e.g. Al_2O_3, CaO increases the specific surface area of the cobalt oxide but the reduction process of this material is slower than pure cobalt oxide [3,4]. Next, very important factor is thermostability of catalytic materials in reaction conditions. The temperature and reaction environment are the parameters that can change the properties of the catalysts. The catalyst deactivation processes are usually investigated by measurements of catalytic activity, specific surface area, the crystallite size or the dispersion of the active phase [5].
The aim of this work is to investigate cobalt nanomaterials doped with manganese, examination of the additives influence on the specific surface area and thermostability under reduction conditions.

EXPERIMENT

The cobalt oxide Co_3O_4, is obtained by cobalt hydroxide precipitation followed by calcinations at 200°C during 2 h in air. Such obtained cobalt oxide, with specific surface area approximately 64 m^2/g, is impregnated by promoters solutions. Precursors of promoters are solutions of calcium, aluminium and potassium nitrates and manganese acetate. Concentration of proper salt is calculated to reach established content of promoters in cobalt nanomaterials. Thereafter calcinations process for 4 h in the temperature at 200°C and for 2 h in the temperature at 500°C is conducted.

Quantitative composition of elements in the materials is determined by atomic emission spectroscopy with inductively coupled plasma ICP-OES (spectrometer Optima 5300 DV, Perkin Elmer). Phase composition is determined by means of X-ray diffraction XRD (X'Pert PRO Philips diffractometer with CuK_{α} radiation). The specific surface area (BET) of the cobalt nanomaterials is determined by nitrogen adsorption-desorption measurements at 77 K using a Micromeritics ASAP 2010 instrument. To evaluate thermostability of these nanomaterials, the specific surface area is determined during hydrogen reduction of cobalt. The change of specific surface area is used as factor to describe thermostability. Samples of each of the catalysts are first reduced in hydrogen at 500°C. Then the specific surface area of the samples is determined using thermal desorption method. Sintering of the reduced materials is carried over in temperatures ranging from 550 to 700°C. At each temperature, the specific surface area is determined at predefined time intervals, until steady value is reached. Details of this procedure are described elsewhere [6,7]. The reduction behaviour of samples is examined by using a temperature-programmed reduction H_2 (H_2-TPR) technique (AutoChem II 2920 Chemisorption Analyze). The temperature increases at a rate of 5 °C/min from 25 to 750°C.

RESULTS

Fig.1 shows the XRD patterns of cobalt oxide (Co_3O_4) nanoparticles obtained after calcination at 200°C and the cobalt nanomaterial with promoters (CoMn0.25) after calcination at 500°C. XRD phase analysis shows that Co_3O_4 is present in all the analyzed materials.

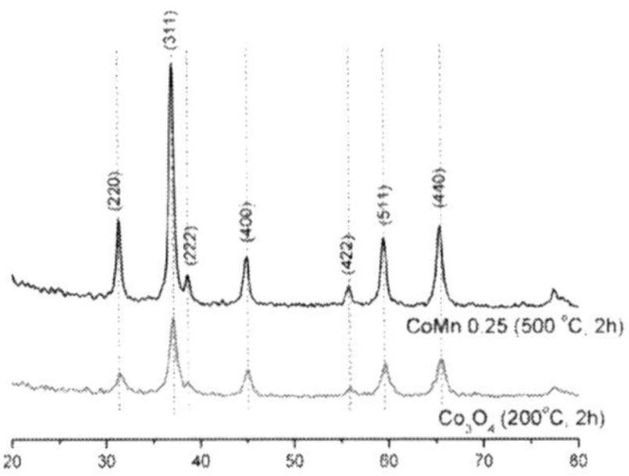

Figure 1. XRD patterns of pour cobalt oxide (Co_3O_4) and the cobalt nanomaterials (CoMn0.25).

The chemical composition and specific surface areas of materials are summarized in Table 1. The quantitative analysis establishes that the content of aluminium, calcium and potassium oxides is almost constant in prepared materials. The content of manganese is only varied. The specific surface area, S_o, of the cobalt nanomaterials increases along with an increase in manganese oxide concentration.

Table1. Chemical composition and specific surface area, S_o, of investigated materials.

Sample	Content [wt %]				S_o
	Al_2O_3	CaO	K_2O	MnO_2	$[m^2/g]$
Co	2.6	1.5	0.5	-	29
CoMn0.12	2.7	1.8	0.6	0.12	31
CoMn0.25	2.8	1.7	0.6	0.25	37
CoMn0.37	2.8	1.7	0.6	0.37	41

Figure 2 displays the variation of the specific surface area of the cobalt nanomaterial (CoMn0.25) as a function of sintering time at each temperature in a reductive atmosphere. The surface area of the cobalt nanomaterial approaches asymptotically to a non-zero value of the residual surface (S_r). An increase of the sintering temperature leads to a decrease of the residual surface.

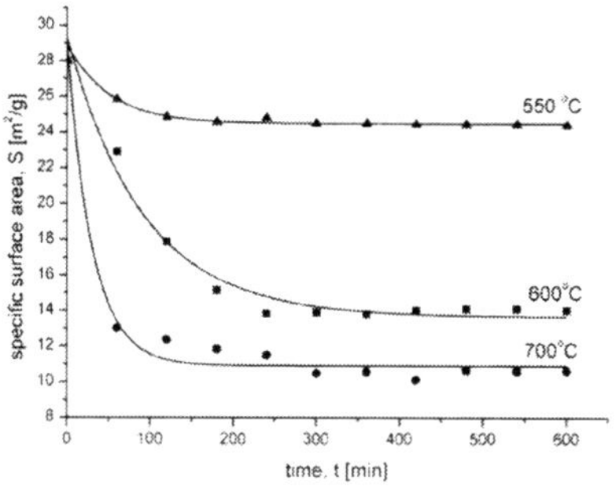

Figure 2. The relationship between specific surface area of the cobalt nanomaterial (CoMn0.25) and the temperature and time of sintering in a reductive atmosphere [5].

Changes of specific surface areas are used as a factor which describes thermostability of catalysts. Examination of thermostability is performed also for other catalysts modified by manganese. In Fig.3 the dependence of $\ln S_r$ as a function of $1/T$ is given. The addition of manganese to the catalyst produces a lower decrease of specific surface area of cobalt catalysts. By using equation 1, the apparent activation energy of the recrystallization process for the investigated materials is calculated [4].

$$\ln S_r = \frac{\Delta E}{RT} + \ln A \qquad (1)$$

where: S_r – the residual specific surface area, E – apparent energy activation of recrystallization process, $R = 8.314 \cdot 10^{-3}$ kJ mole^{-1} K^{-1}, T - temperature (K), A- proportionality constant.

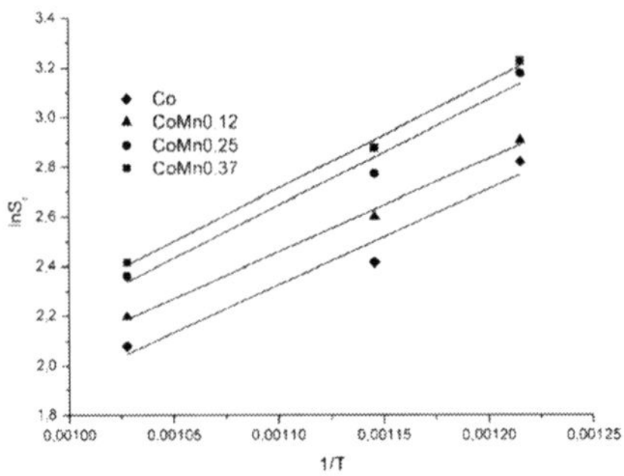

Figure 3. Dependence of the residual specific surface area, S_r, of cobalt nanomaterials on temperature of annealing in an atmosphere of hydrogen.

The values of this energy are calculated as 32 kJ/mol for (Co and CoMn0.12) and 35kJ/mol for CoMn0.25 and CoMn0.37. The obtained values of the activation energy of recrystallization process of cobalt with content of manganese are changing in a narrow range. It can be stated that the amount of manganese, in investigated range, has no influence on the energy of the recrystallization process.

Figure 4 shows the H_2-TPR profiles of cobalt nanomaterials with manganese and without it. The H_2-TPR profiles exhibit two main reduction peaks. The curve of Co has apparent peaks at 294°C and 433°C. The first peak of the H_2-TPR profile can be assigned reduction Co_3O_4 to CoO the second peak is due to CoO to Co°.

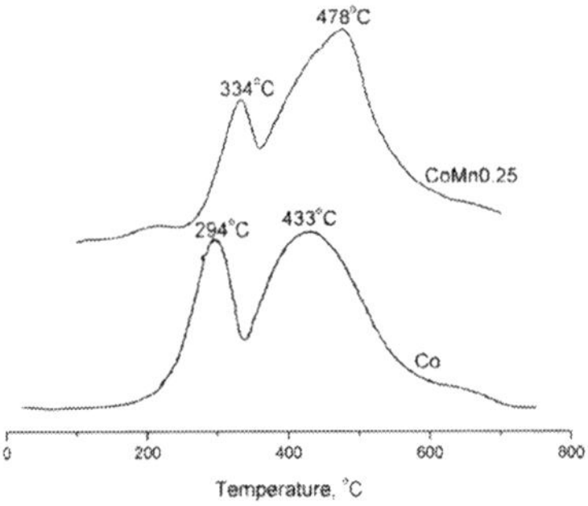

Figure 4 . H_2-TPR profiles.

The cobalt nanomaterials with manganese have begun reduction at a temperature higher than pure cobalt oxide. The H_2-TPR curve of CoMn0.25 has peaks with maxima at 334°C and 478°C.

CONCLUSIONS

The specific surface area of the cobalt nanomaterials increases with an increase in the manganese oxide concentration. Manganese improves thermostability of the cobalt nanomaterials at higher temperature under H_2 atmosphere. Bulk Co_3O_4 is reduced in two steps ($Co^{3+} \rightarrow Co^{2+} \rightarrow Co^{o}$). A small addition of manganese to cobalt oxide shifts the reduction peaks to higher temperatures.

ACKNOWLEDGMENTS

This work is supported by the National Science Centre (Poland).

REFERENCES

1. E. Iglesia, *Applied Catal. A,* **161**, 59 (1997).
2. A. Y. Khodakow, W. Chu, P.Fongarland, *Chem. Rev.*, **107**, 1692 (2001).
3. Z. Lendzion-Bieluń, M. Podsiadły, U. Narkiewicz, W. Arabczyk, *Rev. Adv. Mater. Sci.,* **12**, 145-140 (2006).
4. Z. Lendzion-Bieluń, R. Jędrzejewski, W. Arabczyk, *Cent. Eur. J. Chem.*, vol. 9, **5**, 834-839 (2011).
5. W. Arabczyk, I. Jasińska, Z. Lendzion-Bieluń, *Catal. Today*, **169**, 93-96, (2011).
6. W. Arabczyk, I. Jasińska, K. Lubkowski, Reac. *Kinet. Catal. Lett.*, **83**(2), 385-392, (2001).
7. Z. Lendzion-Bieluń, M. Gleń, *Polish J. Chem. Tech.*, **9**, (4), 5-7, (2007).

AUTHOR INDEX

SUBJECT INDEX

CPSIA information can be obtained at www.ICGtesting.com
Printed in the USA
BVOW011221050412

R4504200001B/R45042PG286324BVX1B/1/P

9 781605 113500